SU
DO
KU

VOL.3

I0490041

MEDIUM
PUZZLES BOOK

Sudoku Rule

Sudoku is logic base puzzle, played on a 9 x 9 square grid. it's all based on the simple principles of using numbers 1 through 9, filling in the blanks and never repeating any number within each square, row or column. The givens numbers cannot be changed.

■ Never repeating any number within each ROW.

	1		5		7			
5	4	2	1	7	8	6	9	3
	3		2		9		8	
9	2		6	8				
6	8						5	1
					2		6	8
	6		7		5		3	
1								
		8		6		5		

■ Never repeating any number within each COLUMN.

		1	5			7	2	
5	4	2	1	7	8	6	9	3
	3		2		9		8	
9	2		6				4	
6	8						5	1
					2		6	8
	6		7		5		3	
1							7	
		8		6		5	1	

■ Never repeating any number within each SQUARE.

		1	5			7	2	4
						6	9	3
	3		2		9	1	8	5
9	2		6					
6	8						5	1
					2		6	8
	6		7		5		3	
1								
		8		6		5		

Tips on Solving Sudoku Puzzles

Tip 1: Try to work in 3x3 squares, rows and columns that are almost full. As you complete rows, columns and squares, you will begin to do it's easier.

Tip 2: work with individual digits. Look to see what digit already have lots of. It can help you figure out the rest of that same digit. Scan up and down, as well as left to right in each row and square. See if you can find more of the same digit. Put one of the digits in each square will start to make the rest of the digits easier to determine as well.

Tip 3: Check back frequently. As you keep discovering new digits scanning every row and column, in all directions and the same with the 3x3 squares and individual digits, and the possible digits marked, find even more correct digits. So repeat the same scans and techniques. frequently until you finish Sudoku.

Tip 4: Never guess a digit if you can't figure it out. Establish a bad digit could make you unable to solve the puzzle and come out with more wrong answers. Just write your options in pencil.

	2	7	4		6	3	5	
								6
							7	
	5		1					4
		1	2		4	5		
8					3		2	
	8							
9								
	6	2	9		7	1	4	

5			3			7		
					1	3		
	1	3			4			6
1				6			2	
	9						8	
	5			4				7
3			2			4	7	
		6	1					
		5			8			1

6							8	3
			5	2		7		
7					3			
				8	6		3	
4				9				8
	5		3	7				
			2					5
		3		6	9			
2	9							6

	1			6	5			
		9	3					
	8			7			6	
		4	7				9	5
6								1
2	9				4	7		
	5			2			8	
					8	9		
			9	4			5	

							8	
2			1	9			5	6
1			7		5			
4						9		
		1	2		4	7		
		3						1
			8		2			4
6	1			3	7			9
	5							

					8		4	
			1		9		6	8
	5							9
		4	3				2	6
		8				1		
3	1				2	4		
1							3	
5	8		6		4			
	7		2					

	2		1	3		9		
	6		2			7	4	
5								
		5	6					9
	8						5	
2					5	3		
								3
	5	8			2		1	
		7		8	3		2	

PUZZLE 8

	9			6	3			
								9
	4				9	3	8	
	1			2	4			5
		9				8		
2			3	8			4	
	3	8	1				2	
1								
			5	3			7	

				4				6
	3					5		2
	4		6				3	
			9	5				8
	8		1		4		6	
5				8	7			
	2				1		5	
8		9					2	
1				3				

	2			4	5			
							2	9
			8	7		1		
2	1				8			7
	7						3	
3			1				5	4
		1		8	9			
8	6							
			4	1			8	

PUZZLE 11

3				4			8	
	5			6	9			
7			1					5
6	7	4						
1								3
						4	1	7
9					7			1
			2	8			7	
	1			5				6

PUZZLE 12

		9	1		3			
	5				8	2		
							8	
1		8	9					6
2			4		1			8
4					5	9		2
	4							
		7	8				9	
			6		7	5		

						5	8	
			4					3
		5			2		7	4
	5				8	9		
6	1						3	2
		4	3				6	
7	3		9			1		
5					3			
	9	1						

				1	3	4		9
	3						7	
			4					2
3		7	5					
5	9						8	1
					6	7		3
4				8				
	6						3	
7		2	9	3				

				4		1	3	
8				2	1		5	
			8					2
					7		4	
9	2						6	1
	4		9					
6				5				
	8		2	3				7
	3	9		1				

	8		9		5		7	
		9	2					
	4			3				
5						2		4
	3	6				8	5	
4		1						7
				5			9	
					8	6		
	9		3		6		8	

8	4							6
		2			4			
					1		2	3
				5		9	1	
1		9				3		2
	5	8		2				
4	8		5					
			6			5		
6							7	4

9					5		1	
	8		7				5	4
								2
5							3	
	3		8		1		2	
	4							7
1								
8	9				3		4	
	7		4					8

					4	8		
		1	3	5			9	
8	9		2					7
	2					9		
6								4
		9					6	
2					3		5	1
	8			4	2	6		
		5	1					

			7	8	6			
		8					1	
9		6		4				
	5			1	3			2
		2				8		
1			6	9			3	
				2		6		5
	1					2		
			9	5	7			

6		2	9		8			5
				3			6	
	4							
	5				7	2		
9			1		3			6
		3	5				4	
							5	
	6			2				
2			6		5	8		1

		5		9			4	
				7				1
3					8	7		
2	7				1	8	6	
	3	8	6				1	2
		7	4					5
8				6				
	5			2		4		

PUZZLE 23

	1	2	7				5	
		8		1				
					6		2	
	3			9		2		
7								4
		5		3			8	
	5		1					
				4		9		
	7				3	6	1	

PUZZLE 24

						1		3
2					5		7	
8			6	3				
							5	8
	3		7	8	9		2	
9	1							
				9	6			2
	5		8					7
4		2						

				2		1		
			8		9			7
	4			5	1			
	8					2	7	
2		3				8		5
	5	6					3	
			5	1			8	
9			3		2			
		2		7				

9		2	5		3	6		8
	7		6					
3				8				
	3	6						
2								1
						7	9	
				6				2
					5		1	
6		7	2		1	5		3

PUZZLE 27

	4				2			7
			1	8		2		
					3			8
		4		7		1	2	
7								6
	3	1		2		7		
4			6					
		6		4	5			
3			8				9	

PUZZLE 28

			9				5	1
		3			8			7
5						3		
6				1	5	7		
	7						2	
		5	8	2				6
		1						8
3			5			9		
9	6				4			

				4	1		7	
3	4							
	8			3		5		
	5				3			6
7			1		4			8
6			7				4	
		6		5			8	
							3	5
	9		8	2				

2			8					
	5			1		9		
7		8				2		3
8					5	7		
			7		3			
		2	1					6
6		9				8		2
		3		9			6	
					1			9

PUZZLE 31

		1	8		4			
				1		4		
	8		5	2			3	
							8	2
2		3				5		7
1	9							
	3			4	2		6	
		4		7				
			6		5	8		

PUZZLE 32

5					4	3	9	
	3			8			5	
	2							1
			7		6	5		
		2				4		
		9	5		1			
3							6	
	1			3			7	
	4	5	9					2

1			4	8				
		4		5				2
	5				1		7	
		6	7				3	
4								7
	3				5	1		
	4		1				5	
3				9		6		
				4	6			9

6					1		9	
	5	3						7
				5				
2				4		9		
9	8		5		7		2	1
		1		2				8
				8				
5						4	8	
	7		6					9

PUZZLE 35

9					7	4		
			6			1		
	4		8					5
					4		2	7
	7	1				5	8	
6	8		5					
5					6		7	
		4			8			
		9	7					4

PUZZLE 36

5							4	7
	1				7	5		6
8								
		8	9				7	
		9	6		3	2		
	7				1	6		
								8
9		5	7				6	
3	6							1

	8			9			3	
					2		8	
2			1					6
		5			9			7
	1		5		4		6	
8			2			5		
1					5			4
	3		9					
	9			6			7	

8	1		3		4			
							9	
		5		6				4
	8		9		2		5	
3								8
	5		8		1		4	
6				4		2		
	7							
			7		9		3	5

			2				4	9
				5			7	3
			4		3	2		
7	8						1	
			7		2			
	2						8	5
		3	5		8			
1	7			3				
9	4				1			

2		3	7					9
6						7	4	
				8	3			
			4	5	7		9	
	4		2	8	9			
		7	6					
	8	1						2
9					3	4		1

	2		6					9
4				3	7		5	8
		7						
	6				1		9	
3								4
	7		8				6	
						8		
8	3		2	7				5
6					4		1	

		4			9	3	6	
				3	7		4	9
1				5				
2	1					5		
		6					9	3
				8				1
6	8		5	2				
	3	5	9			2		

PUZZLE 43

			3	7			1	6
							7	
		7	6		9	3		
	3				5			
1	4						9	3
		8					4	
		2	4		6	8		
	5							
6	7			3	8			

PUZZLE 44

	1	8		4	7			
7					9		1	
				2		7		
		5					3	9
1								4
3	2					5		
		4		8				
	3		9					7
			3	1		4	5	

5			4			3	9	
				3			4	
		3			5			1
			8			7		2
	4						6	
8		7			1			
7			5			4		
	3			9				
	9	4			3			6

9		4	7	2			1	
1								
		2		6				9
	2		9					8
		3				6		
5					3		7	
3				1		5		
								7
	7			5	8	1		3

PUZZLE 47

	7			9			8	5
	4		5					
	6		7			1		4
		6						9
	5						7	
8						3		
4		1			6		9	
					8		5	
5	8			2			1	

PUZZLE 48

1					8		7	
	4	3		2				
5					1			
					7	4		
		7	9	5	4	3		
		6	2					
			5					2
				4		5	6	
	2		3					8

		3	4	5				8
		7			1		5	
						2	3	
				2		6	4	7
6	8	4		1				
	3	8						
	6		8			9		
1				7	3	8		

	3					7		
7			9			2	1	
		9		4				5
		4	3					
			7	1	9			
					4	1		
3				5		9		
	9	5			6			2
		8					6	

PUZZLE 51

		9			8		1	
5		6		2				
	4		7			2		
2					4			8
1								2
6			3					4
		2			9		6	
				3		4		9
	3		4			5		

PUZZLE 52

			2			6		
				1		7		9
	8						2	
7					5			8
6		5	3		8	1		7
9			1					2
	6						5	
3		2		6				
		9			4			

PUZZLE 53

	9			6				
			5			3		
2		5	4					
		2			4		1	9
9	8						7	5
5	1		9			6		
					6	1		2
		3			2			
				1			8	

PUZZLE 54

		8			1	4		
5				9			6	
			8			1		
4		3	2					9
7								4
9					4	8		3
		4			8			
	3			5				1
		5	6			7		

PUZZLE 55

		3	4	8				
5		7			9		2	
8	2						1	
		1	5		3			
			6		4	9		
	7						6	2
	6		2			3		9
				6	8	7		

PUZZLE 56

		4			5		7	
				4				6
2					1	4	5	
				6				2
5		3				7		4
9			3					
	4	9	1					5
1				8				
	7		9			1		

1			6		9	5		
		4				3		
				5				1
	6				4			9
	2		3		8		5	
4			1				7	
8			7					
		7				1		
		9	5		2			4

7					9	3	5	
					3	2		9
	5			8		1		
	8		7					
					4		3	
		2		4			7	
6		9	8					
	7	5	1					4

PUZZLE 59

2				5	3		1	
	4			9				
		8			7	3		
4	8						7	
		6				9		
	1						4	2
		7	8			1		
				1			9	
	2		5	7				4

PUZZLE 60

		9	6		1			
		4					3	
5				9		8		
	4	5						
	6	1		7		9	4	
						5	7	
		2		5				3
	9					4		
			3		4	6		

			8	5			1	4
		5						
7					1	9		
4				3		1		
	1		2		6		9	
		8		9				6
		2	5					9
						3		
6	7			2	3			

		7	4					
			2	3				
2	8							5
1			9	6				8
	6	8				5	9	
7				8	2			6
8							2	7
				1	4			
				9	6			

		1						
8	3			6	4	2		5
	6				9			7
7				5				
		3				6		
				4				9
1			6				9	
3		5	9	1			8	6
						3		

	4			6				
		1				6		
7	2				5	8		
			3		9			1
		4		1		7		
6			8		7			
		9	2				1	8
		8				2		
				3			4	

3				4				
2	4				3			
	5		1		9			3
					1	8	9	
		3				6		
	1	8	5					
4			9		7		1	
			8				6	4
			3					8

					3		9	4
		3	9			1		
	1			8	6		5	
	5			3				6
7				1			8	
	3		6	4			1	
		6			1	2		
1	8		2					

PUZZLE 67

				4				
		5	3				1	
	1				5	8	2	4
		4					6	
	3		1		6		5	
	5					7		
1	6	7	5				8	
	8				3	5		
				7				

PUZZLE 68

		4			9		3	6
		9			4			
	7		6			5		
9			7			6		
4								2
		1			5			8
		3			7		8	
			8			3		
2	8		9			7		

		9				2		1
		3		1	2			4
			8				3	
			1	6				9
	4						8	
6				3	8			
	8				3			
7			2	5		4		
4		2				5		

		4						5
5	1		2					
8			3			7	2	
				6			9	3
			7					
2	9		8					
	5	8			3			9
					5		8	4
9						1		

			9		8			5
	2		6	7				
5						6	8	
		1	4					
	9	7				4	3	
					2	7		
	8	5						6
				1	9		4	
4			5		3			

4					9			8
	8		6				2	
				4		1		5
					1	8		3
			2		4			
9		3	7					
8		9		7				
	5				2		9	
2			5					1

			1	9				4
				8		6		1
		6	4				7	
	6					5		
	7		9		8		1	
		4					3	
	9				4	8		
5		1		3				
8				5	9			

	6				5		9	1
		7					8	
	3		4			5		
	2			7				
		1		8		2		
				3			5	
		5			3		4	
	4					8		
9	8		6				1	

PUZZLE 75

			9				4	6
2			1					
						7		5
		5		2	8			7
	6		4		5		8	
3			7	6		1		
9		6						
					2			8
5	8				4			

PUZZLE 76

	6	4	2				8	
		2		7	9			
				1	6		2	
		1				7		
	2						5	
		8				1		
	3		6	2				
			7	3		4		
	4				8	2	6	

	6		3					
			2			1		
		5			7		3	
	5		4			9		2
6								5
7		8			6		1	
	4		5			7		
		3			2			
					9		8	

		5			1	9		
	6		8	9	3			
								6
3					9	2		
7	9						4	3
		2	6					9
2								
			4	1	7		8	
		4	9			1		

2				3				
		4	5	7			9	
	8				1			
	1	2			5			8
		3				9		
9			1			2	3	
			2				4	
	5			4	6	3		
			5					1

			4				2	3
	5	6			9			
2			6		7			
		3						1
		7	3		1	4		
9					2			
			9		4			8
			7			9	6	
6	7				2			

							2	8
			3	4				
		9				3		4
	8			1		2		5
1				4				6
6		2		7			8	
5		3			6			
			8	9				
2	7							

	6			5			1	
				7				
3				8				5
8		1						
	4	5	6		3	8	7	
						3		1
5				4				9
			9					
	2			7			5	

PUZZLE 83

		4	7	2		6		
	7				5			
1				4				
2		7			1			
8	4						9	1
			6			3		7
				9				2
			1				4	
		2		3	7	1		

PUZZLE 84

	1	7			9	6	5	
4								
			7				1	
2				4		8	9	
			2		3			
	5	1		8				4
	4			3				
								9
	2	3	7			1	6	

	3	6			9	4		8
	5				7			
			2					9
8							4	
		4	7		8	6		
	9							1
7					4			
			1				8	
3		1	8			7	6	

		5	8					
			3	7				5
	2					9		
	1				6			8
	6	9		4		5	7	
8			9				3	
		3					8	
4				5	1			
					3	6		

PUZZLE 87

5			6					2
	6			2				
				3	7	5		
	8				3		2	
6		1				3		8
	4		8				7	
		4	7	9				
				4			5	
3					5			1

PUZZLE 88

8		3				1	9	
			7					5
					8	7		
	4		8	1	6			
		6				5		
			3	5	4		2	
		8	4					
1					5			
	6	5				8		3

				7				2
9		7				6		
		5		1		4		
	3				7			6
	7			9			3	
5			2				1	
		3		4		1		
		6				3		5
8				5				

	9	4	1	3				5
	6							
3			9				2	
		9					4	
		1	4		7	2		
	2					1		
	7				2			4
							1	
6				9	1	8	3	

							2	
		5	3	4				
	1	8			2	3		
9	6						1	
	7		2		9		5	
	4						7	9
		6	8			1	9	
				2	7	6		
	3							

	4				8			1
						3		
	1		3	7		9		
				9				8
	9	1		7		4	3	
3			4					
	6		9	1			2	
		7						
8			7				4	

		3	8				5	1
							3	2
1				3				
			5	7		9		
		5	1		8	2		
	2		3	4				
			2					4
2	9							
7	6				5	8		

2					7		4	
							9	
1		5		8				
4	1			7	9			
	8			1			6	
			8	3			1	4
				2		9		3
	5							
	9		1					2

PUZZLE 95

		8					5	
9				2				6
5	1		7					2
	3	5						
	8		3		5		1	
						9	3	
8					2		6	7
4				8				9
	7					8		

PUZZLE 96

4				5				8
			3			7		
9					6	3		
		3	2	1				
		9				2		
				7	9	6		
		7	9					6
		8			3			
1				8				5

6					3			7
9					1	3		
	8	3	7					
			5	4		2		
		2				8		
		1		7	2			
					8	7	3	
		8	9					4
5			1					9

4					6			1
	8		4				7	
			8	5	9			
8	2							
5	4						8	9
							4	3
			5	6	3			
	9				4		2	
3			1					4

	9					6		
			4		5			
	4	5	3	2				8
		3			8			4
	7						1	
5			2			8		
1				5	4	2	8	
			8		9			
		4					7	

7			5					
	4	5			9			6
			2				4	
		4	9			7		8
				1				
8		1			3	4		
	9				5			
3			4			6	8	
					6			1

		3		1				
6	5		3			1	2	
				5				7
	4							
	1	5	9		7	4	6	
							8	
5			2					
	8	9			1		5	3
			7		6			

					9	5	6	
		4			8			1
	1		7	6	2			
1	7							
		8				7		
							3	6
			2	7	3		9	
3			9			4		
	8	2	5					

PUZZLE 103

		9				7		
5	8			1				9
			5					2
	7	6			5	8		
8								1
		4	8			2	3	
4					2			
7				5			4	3
		3				9		

PUZZLE 104

		6	2		1			
		1	5		9			
7						2		
		4	1					7
2	8						1	9
6					8	4		
		9						6
			8		6	1		
			9		3	8		

6	2		4					
4		1	2		3		9	
					5		8	3
2			1		8			9
7	8		6					
	7		3		2	8		1
					9		3	4

3			7					6
	6						8	
			4	1		5		
4		5			7	6		
2								9
		7	1			3		5
		6		3	4			
	2						6	
7					5			2

PUZZLE 107

1				6	4			3
			5				2	
7						5		6
9	7			1		8		
		3		8			1	4
2		8						1
	1				7			
6			2	5				8

PUZZLE 108

		2	7	4				
				3	1			
	6					8		
1	4			2			8	
7		6				2		9
	5			9			7	1
		3					2	
			9	7				
				1	5	9		

					5	8	6	
			2				7	4
			6			9		
		5		7	2			
3			9		6			2
			3	1		5		
	7			8				
6	8				7			
	5	4	6					

	8			7			6	
		6	3			9		
1	7							
				9		1		8
	3		7		1		4	
5		1		4				
							2	6
		3			5	4		
	2			6			5	

9		1			4			2
	4		7		6		1	
						4		5
		7	9		1	3		
5		3						
	6		4		2		7	
2			1			6		3

	6			9				
1		2		7				
		4	5			1		
4			8	3			7	
	7			6	5			3
		9			4	2		
			1			3		9
			3				5	

	8		5			7		6
9					3			
		6			7		1	
	3		4					8
			2		6			
1					5		2	
	2		6			8		
			3					4
8		7			2		5	

7	5			3			6	
						2		
	1		8			3		
			2		3	1		
9								6
		3	9		8			
		5			7		4	
		9						
	2			4			7	3

PUZZLE 115

				3			7	
		9						6
	9	4			2		3	
		9					2	
	3	6		4		7	1	
	2					5		
	5		8			3	6	
7					6			
	8			5				

PUZZLE 116

				9				
	9				4		1	
4		7		6		5		
		8				4		2
	4		8		5		3	
3		6				1		
		3		2		8		7
	1		9				5	
				8				

2			8		5			
7				9	1			3
			3				4	
	5	6		1				
4								2
				4		5	8	
	2				9			
9			2	3				8
			1		6			7

4					6			
7			9			8		3
			3				7	
		4			9	7		
8	7						4	1
		9	5			6		
	1				5			
2		6			7			4
			2				9	

9			2	8			7	
7				1		2		5
				4				
	1					9	2	
2								1
	9	8					6	
			7					
4		3	5					6
	8		9	1				3

			7	6	3		4	
		7					8	
5			4					6
7					5	9		
			8		6			
		4	2					5
3					1			4
	6					1		
	9		5	3	8			

9	2						3	8
7					3			
	4		6			2		
8	3			6				
				5				
				2			9	4
		1			6		2	
		8						1
3	9						4	5

	8		4	2				9
						2	3	
7		5	3					
	4		9			7		
2								6
		1			6		4	
					7	3		4
	1	6						
9				8	3		6	

5	6							
8			3	5				
	7					5	9	
				3		4	1	
9			6		7			8
	8	4		9				
	5	7					6	
			8	5				3
							5	1

	2			8	1		6	
6			2			8		
	5						3	
9				4	3			
			9		6			
			1	2				9
	8						4	
		7			8			5
	3		6	1			8	

	6							3
7			4			8		
3		9			2		1	
	8	4						1
			5		7			
1						6	2	
	3		1			5		7
		1			4			2
9							3	

								3
	1	3	4	2				
		5		3	9			2
				9			8	
4		8				1		7
	3			7				
8			2	6		7		
				4	8	5	9	
5								

								2
	7					4		
	1		6	5			3	7
1					4			
4		8		9		3		5
			5					6
2	5			8	1		9	
		9					7	
7								

2	6	4						
							1	
					6	2	9	
		5		6	2			8
7			5		4			9
3			7	9		1		
	7	3	2					
	8							
						6	3	1

	5					7	2	
	3			7				9
9	4		1					
					8			1
	9	4				2	6	
1			9					
					6		5	2
6				4			7	
	7	9					1	

	1		8		4			9
	5	6	2			1		
			6					
		4			2			6
9								7
8			7			2		
				2				
		9			7	6	4	
3			1		6		5	

PUZZLE 131

6					3			
9			4					
	1	7	8	5				9
1		2					3	
				7				
	9					4		5
2				1	7	9	5	
					5			7
			3					1

PUZZLE 132

	6					4		
			9					5
4			5			1	3	
	9		8	6				2
				3				
8				1	5		7	
	5	4			7			8
2					3			
		7					2	

PUZZLE 133

	8			6		2		
	3	7				4		
	4		9					
	6	1		8		7		
3								4
		4		3		8	5	
					6		4	
		2				3	1	
		9		1			7	

PUZZLE 134

6	8				1			
3		1					5	
			6			1		3
			7			9		
			5		2			
		8		3				
1		3		5				
	7					4		2
			2				3	1

	2			5			6	
	8							
	6			3				
2	6				5	7		
5		4		6		9		2
	7	8					5	3
		7				8		
						1		
	1			2			4	

9			5					7
4							5	
1					2	3		
		9	4	5				
		5		9		2		
			2	6	8			
		4	6					2
	3							1
2					3			8

			5			1		
9	4		3					
					9		2	7
3	9						6	
	5			4			1	
	6						7	8
6	2		9					
					6		3	1
		4			8			

	9			2	7	6		
		6						3
2						8		
		3			2		5	8
			8		3			
1	8		5			2		
		5						6
7						1		
		4	9	7			8	

		6	7		1			8
		2					6	
7					9			
8	7			1			9	
		1				2		
	5			4			8	6
			6					9
	1					6		
3			1		5	4		

	6	4						
	5				7			1
			4		5			2
	4				8		2	7
		2				3		
5	9		7				6	
8			9		2			
7			5				8	
						2	1	

		5				1		
1	6		7				5	8
		8						2
4					9		1	
				2				
	7		1					9
5						7		
9	1				8		6	3
		3				2		

7				8		2		
2		4						
	3		1	2	7			
	5				4		1	
		7				5		
	8		3				7	
			5	3	9		4	
						3		8
		5		6				1

			6	1			7	
					8		2	
	9			2		3		
6		5			3			8
3								2
4			9			5		7
		2		7			9	
	5		8					
	4			6	2			

	4			6	8		3	
	9	3						
		8			2			9
	6			8				5
		7				6		
2				3			1	
8			4			7		
						3	9	
	3		2	5			6	

		1	4	5				
	3				9			
4			7					1
7		2						
6		5	2		3	9		8
						5		6
1					7			4
			8				3	
				4	2	8		

		7	9					8
		1	6		5			
8							1	
5				7				4
7		4				2		6
3			5					7
	8							5
			7		3	8		
4					2	3		

PUZZLE 147

	2	7	6					
				3				6
8				7		3		
		8			6		1	
5	1						7	3
	9		1			5		
		3		6				8
1				2				
					4	9	3	

PUZZLE 148

	4			5	8	9	3	
	9			1				6
					9			5
	3				7			
		1				5		
			4				6	
6			5					
7				2			8	
	2	9	8	4			5	

			4	8			5	
1				7		3		
		6	2					
9						2	3	
3		1				5		9
	6	2						7
					8	9		
		4		3				8
	3			6	4			

	2						1	8
			7		6	5		
8						7		
2				3				
	5	1	4		8	3	6	
				1				2
		4						5
		9	8		1			
1	8						4	

PUZZLE 151

			3			7		
	3				8			2
				2		5	3	4
4	5					9		
			8		4			
		7					5	6
7	4	1		8				
2			7				1	
		6			9			

PUZZLE 152

9	5		7		1		8	
			6					5
2							3	
				1			6	
1			5		3			4
	9			2				
	1							3
3					8			
	8		3		5		9	2

		1		6				
	6				2			
8		5	9				3	
5			7				9	
		4				3		
	9				8			5
	2				7	6		3
			3				5	
				9		4		

	2				8			
		1		3				4
9	8		1			3		
				6	2	5		
	3						7	
		2	3	8				
		7			6		2	1
8				1		6		
			8				9	

PUZZLE 155

			7					
		4				8	6	
2		7		4		5		
6					1			
5	1		2		7		8	6
			8					5
		6		3		9		8
	9	2				3		
					2			

PUZZLE 156

	8				1			
		3	5		6	9		
	1		4		9			8
			3			6		
2								7
		6			5			
5			2		7		4	
		7	9		3	2		
			6				5	

	6	3		9				
				4	1	6		
8			3			9		
	9							4
	8		2		4		5	
7							2	
		6			9			7
		5	6	8				
				7		1	6	

			8			3		
6	8			3			2	7
				9		6		
		1	7				9	
	2						3	
	6				1	5		
		3		8				
7	5			6			4	8
		6			4			

4							1	
			3	7		4		
					4	3		8
3				2		8		
	6		1		3		7	
		7		9				2
1		6	9					
		9		5	8			
	4							7

				5				9
	3						6	
6					8	1		
				2			3	4
		9	1		7	5		
5	4			3				
		1	9					2
	8						4	
9				8				

			9		6		5	
8				1				7
		1	5					3
		7						6
		4		5		7		
6						4		
1					3	9		
3				7				1
	6		8		1			

				1				4
6		7		5	4			8
				6			9	
	4					2		5
		3				8		
7		2					6	
	9		6					
8			4	9		1		3
1				7				

		5	3				4	8
3			1	4				
	8					7		
		2		1				
1	4						6	9
				5		3		
		3					7	
				9	2			3
2	7				5	8		

	4	7	6					
				4				9
		2	3		5		4	1
5						6		
9								2
	2							5
7	3		9		4	8		
6				3				
					1	3	5	

1	2			6				
	3				1	5		
	9					4		
4	7	6			2			
			5		3			
			4			8	2	7
		3					8	
		2	8				5	
				2			9	6

	7		2				4	
	2				9		3	
6		8						
2	3				6			
1				8				9
			7				5	1
						8		4
	6		3				9	
	9				7		2	

	5					1	6	
		3		4		7		
				9	7		2	
4	6	2	8					
					2	9	8	6
	4		2	8				
		5		7		4		
	8	6					5	

3		6	8			5	4	
			2	7				
		7			6			9
4						6		
	9						1	
		5						3
5			4			1		
				9	7			
	4	9			5	7		2

			3		7		4	6
9				2		5		
	8							
6	5				9		7	
		8				9		
	4		7				6	2
							5	
		6		4				3
8	7		5		6			

	8	7			5			
5			2					7
				9		3		
		3	4				9	
	2	1				6	7	
	9				8	2		
		8		3				
7					9			4
			8			7	2	

PUZZLE 171

	5		3			8		
6		4						5
	1			2	9			
				5				2
	4	2				6	8	
8				7				
			9	6			3	
4						7		9
		7			1		5	

PUZZLE 172

	8				6			
6		9						7
		3	1	2	5			
					2		8	6
		2				9		
8	6		4					
			5	8	4	1		
4						8		9
			2				4	

4	8				3			
6			4	5		7		
		9						
				7	2	6		9
3								4
5		2	9	8				
						8		
		1		4	7			6
			5				7	3

	8	4					5	
2								
5				3		2		9
	4				7			
9			3	2	5			4
			4				1	
4		3		9				7
								2
	7					6	4	

PUZZLE 175

			8					1
						6		5
5	8		6			9		
		1		5		4	3	
6								7
	5	7		3		1		
		5			7		9	2
9		2						
4					2			

PUZZLE 176

1	2		3		9		7	
				8		6		
	5			2	1			
2		8						
			8		4			
						7		1
			4	5			9	
		4		9				
	7		2		8		4	3

9					2			
		1						3
7	3				5		9	8
				8		4		7
			3		6			
6		8		5				
8	5		6				3	9
3						6		
			7					1

3	5				8	6		
			9				5	3
		7			2			
	4				7			9
2								7
8			5				2	
			8			3		
6	8				9			
		5	1				7	8

					5	1		
		7	1			6		2
2		8		6				
	3				1			
1		9				4		5
			7				6	
				8		5		9
5		4			6	8		
		2	5					

	3		7					
7	4					1		8
		5	9				7	
			6				3	5
		4				2		
6	2				3			
	1				9	8		
5		9					4	7
					4		5	

5					3	4		
7				5				
	4			7	8		5	
	5	7						2
1								4
2						9	8	
	9		7	3			4	
				8				1
		8	2					3

		9				7		
2	8		1					9
7	3				8			6
			4	9				
9								3
				3	5			
6			8				7	2
3					4		5	8
		2				4		

PUZZLE 183

8			5					
6		7				3	4	
		2						5
				3				7
		4	2		7	6		
1				5				
4						8		
	9	6				1		2
					6			3

PUZZLE 184

		4			6	2		
				5			9	
3					2			5
	4				1			9
		2	6		4	8		
9			2				4	
4			3					7
	9			2				
		6	8			1		

		3					6	9
6			4					
	9			8			2	
7	4			2		3		
		8				1		
		2		9			8	4
	7			6			3	
					7			8
1	5					6		

					8			1
	5					9		
		7	3				8	2
2	3			6				5
			2		9			
9				4			1	3
6	4				2	7		
		9					6	
5			9					

3	5	9						7
		2		6	7	5	9	
			2					
	1	5	8					
					1	4	5	
					2			
	6	8	7	1		9		
9						1	7	4

	3		5					
		9		3	2	5		
7			8				3	
		3		5	8			
4								2
			7	2		6		
	5				4			9
		6	9	7		3		
					6		1	

			3				6	
				2	8	9		1
		5	9					8
4	8					5		
				3				
		2					8	9
2					9	7		
7		9	1	4				
	3				5			

			8		1			5
8				5				
9						6		
	9	5	6					3
		6	5		7	9		
7					3	1	5	
		2						4
				3				8
5			7		8			

	7		8			6		
5		9		2				
	8			1				3
	2		7					9
		4				2		
1					2		6	
4				7			9	
				3		7		4
		6			5		1	

			7	5				
	4	9				2		
2					1			3
3								
6	2	5				4	1	7
								6
4			6					2
		3				5	8	
				1	8			

	1			2			7	
			5			3		
	7	8			1			4
				9		4		2
	8						6	
5		2		1				
1			6			7	4	
		9			2			
	6			4			2	

					8			1
						8	7	
3				6			4	
		3			5			
4	2		1		3		6	9
			6			2		
	7			9				8
	4	1						
6			7					

6	7		2					
	9			8		5		
		3	4		9	7		
4	1						7	
	5						8	4
		7	6		8	3		
		1		3			5	
					7		6	2

				5	4		9	
7					1			
5				3		7		
4			8	2		5		
		6				2		
		2		6	5			3
		8		1				2
			3					6
	1		4	8				

	6						7	
		4	1		9	2		
7					3			5
	4				2			
1			4		7			9
			5				4	
4			7					3
		1	9		8	4		
	3						5	

	6			3				
9			8					
	1				9			2
6		7		2		8		
	3	4				2	7	
		5		4		3		9
7			3				4	
					1			6
				6			9	

PUZZLE 199

4		7			8			
1	2		4					
		8		1		5		
7						4		
	1	3				7	8	
		2						3
		9		3		1		
					6		3	7
			5			6		2

PUZZLE 200

2			8		7			
	8			3			2	
		3		6				8
		7	4					
	9	8				7	5	
					9	3		
9				1		6		
	5			9			8	
			6		5			4

5				2				9
2			1	8				
	9				5			
		9					6	5
8	2						4	3
7	5					1		
			4				9	
				5	7			2
4				6				7

5	2			6				
		6	1	5	4			
	4				7			
		4						6
1	3						5	4
8						3		
			8				3	
			9	4	3	8		
				7			4	1

7					1	9		
8				2		7		
		4	8					6
	5	6					1	
			5		9			
	1					4	3	
5					4	2		
		2		7				4
		8	2					7

	6	7				9		
					3			6
		9		8		4		1
4				7			2	
		1				8		
	3			1				4
1		8		9		3		
5			1					
		3				7	1	

	3				2			8
5	2							
		4		6				
1		5		9		6	4	
				5				
	6	2		8		5		7
				4		9		
							1	2
4			5				3	

	5	9			6		1	
8				5				
	1		7			4		
						2	9	4
		1				8		
6	9	2						
		5			3		7	
				9				6
	7		2			3	4	

PUZZLE 207

			6		8		9	
9		4		3		6		
	8							
				8	6			9
	3	6				4	5	
2			1	4				
							1	
		3		7		9		6
	7		8		1			

PUZZLE 208

		4	7					
	9					8	7	
			3	9	5			
6			2					8
	5		6		9		3	
2					1			6
			4	6	3			
	1	3					8	
					7	9		

				9	6		3	7
2	9		5					
		4		8		6		
	3							8
		1				5		
6							2	
		5		6		9		
					3		8	5
9	4		8	7				

	9				5			
	6							1
		5		3		2	6	
	2		6			1	7	
			4		8			
	1	6			3		4	
	5	7		9		4		
1							9	
			3				5	

PUZZLE 211

	7	2		9	4			3
			5				9	4
		9						
1			8				6	
		6				3		
	8				6			2
						1		
8	5				1			
6			3	8		4	5	

PUZZLE 212

			4					
	8			9			6	
		5	6	1		2		
	3	6						8
		4	5		9	7		
9						5	2	
		7		2	6	1		
	1			8			9	
					4			

			5					8
				7		6		
8					1	2		
	6	5					8	7
			8		4			
2	8					1	6	
		3	1					5
		6		5				
9					2			

	7		8					9
				1	3	4		
	1		7					
1						2	9	
3		5				1		6
	8	2						5
					1		4	
		1	6	9				
2					7		3	

PUZZLE 215

	8		4			5		9
6						7		
				9			3	
		1		4	6			
	9			2			4	
			9	7		2		
	1			5				
		6						8
3		7			1		5	

PUZZLE 216

	7	2	5		3	8		
	9					3		
		4			9			
	8			2		9		
				6				
		7		3			5	
			7			2		
		8					6	
		6	4		8	7	1	

3								5
			8	1	5	9		
			9				6	
		5		2	9		1	
1								9
	4		3	6		8		
	6				4			
		9	6	3	8			
5								1

	1	9		8			2	3
		6					7	
						4		
2				5	7		8	
			6		4			
	5		8	1				9
		3						
	7					5		
9	8			4		1	3	

8				1			2	
	1		7	8		5		
	6				9			
		2				7	4	
			9		5			
	3	6				8		
			1				8	
		8		4	7		3	
	9			5				6

9			4				6	
					2			8
2			5	9		3		
		1					2	5
			7		6			
7	3					6		
		6		3	5			2
4			1					
	8				4			3

5				4				
	2				9	3		
8	1		5					7
	6					9		
			6	2	4			
		3					6	
4					3		8	1
		5	2				9	
				6				5

	3		4			1		
4								5
			2	5	8		9	
				2				6
	9		8		3		2	
5				9				
	4		7	3	6			
9								2
		3			2		7	

					7			
		4	2	3			6	
3			6				8	
6		7	8					
1				2				8
					6	1		5
	9				5			2
	1			9	8	5		
			7					

		2			4			
5	6				2			
			5				3	1
		4			8	7		
6	7						1	8
		5	1			6		
9	8			7				
			2				9	4
			9			1		

5		6		8			9	
		8		2	9	6		
		2			3			8
9			8		6			5
8			2			9		
		1	9	3		8		
	7			6		2		3

					1			3
2			3		8			5
		8	5			6		
		7				2	9	1
1	6	4				5		
		5			9	8		
9			8		2			4
4			1					

PUZZLE 227

9			7					
6		2		1		5		
					2			3
		1			3	4		9
			1		7			
2		4	9			1		
4			2					
		5		8		3		2
					9			6

PUZZLE 228

					8			
	4			5			9	8
	6		4	9		3		
					6			5
3		9				7		6
5			3					
		5		7	2		4	
8	2			1			6	
			8					

		8			3		6	
3		6		9				
		1						5
	7		5		6		2	
				1				
	8		2		7		4	
5						2		
				2		4		1
	2		8			7		

		3		1				7
		7					8	
	9				7	1		3
2				3			6	
			5		8			
	4			7				5
5		6	9				1	
	1					3		
9				8		4		

		1	4	8	6			
	3						8	
		7		3				
6		4	1				7	
			3		7			
	7				5	1		6
				5		2		
	8						9	
			6	2	4	3		

		7		8	1	9		
				6				2
	9	8			2			4
	5	2						
			3		6			
						4	5	
6			8			3	7	
7				2				
		4	6	9		5		

7	1		3					
3		5		2		4		
					7			
		2		8		7		5
	3						4	
5		4		9		1		
		9						
		1		3		5		9
					1		7	6

	3						5	
4				9		7		
7				1				2
9	4				8		6	
		2				4		
	8		4				3	9
3				8				1
		6		5				8
	1						7	

PUZZLE 235

		3		2	4		5	
				9			6	2
					6			
	9	1						7
		5				6		
3						4	2	
			4					
2	7			8				
	8		3	1		7		

PUZZLE 236

	2	1				7		
		5	3					
	3		9		2		4	
					6			
	7	6		8		5	3	
			2					
	8		5		9		7	
					7	6		
		7				8	5	

			3		6	7		
		1	4			6		
	6						5	1
	5	4						3
			2		1			
3						2	9	
5	4						7	
		3			2	1		
		2	5		7			

		6	9	4			1	
4	7							
		9	8				3	
			5					8
	4		7		3		9	
6					8			
	1				4	7		
							6	9
	6			8	2	1		

PUZZLE 239

	8	5			2	9	1	
							6	
		3	1					5
		2		6				7
			9		4			
4				3		8		
8					7	6		
	5							
	9	4	2			7	5	

PUZZLE 240

9	4							
1		5				3	9	
			8	9	2			
	7		3					9
		1		7				
6			9			7		
		3	2	4				
	5	4				8		3
							4	1

			9	2			5	
		6				3		
9								8
3		2		1		5		
	9						8	
		8		3		2		4
4								5
		1				6		
	8			7	3			

3		8						
	2		4					1
	1		7			2		
	3		6	9		1		
	9						2	
		5		1	8		3	
		4			1		7	
6					4		1	
						5		2

PUZZLE 243

			9	8				3
	4		1					
5	6					1		
	3		2				1	
2				1				7
	7				4		2	
		9					5	2
					8		3	
6				9	7			

PUZZLE 244

		4			6			2
5	9		3			7	4	
9	3			2				
6		2				9		4
				6			3	8
	8	9			4		5	1
7			5			8		

5					6			4
7		3		2			1	
2	6							
	5			1				9
			6		5			
1				7			8	
							4	5
	7			5		8		6
3			9					2

		8	4	2			7	
	1							
		3	6				9	
			8	7				5
1								4
5				9	2			
	2				5	1		
							2	
	4			1	9	6		

PUZZLE 247

							3	5
6		2	1					
	7		8			4		
4	8			5				
9	6						5	3
				3			8	4
		8			6		9	
					7	5		1
5	3							

PUZZLE 248

	1		5					
3				2		6		
2		8			4	5		
			6		2	9		
	6						5	
		5	3		7			
		1	8			3		4
		4		3				7
					1		9	

3		9			2		1	
	2		1		6			7
			9			8		
	3					9	5	
	8	7					4	
		1		4				
8			6		9		2	
	7		8			1		4

4				8			1	
							3	6
				3	7	5		2
6						2	7	
			5		4			
	8	7						1
5		6	8	9				
8	2							
	7			2				8

	6		2			3		
7								
	5		9		3			4
					9		8	1
		5	3		4	2		
6	4		8					
2			1		6		4	
								9
		3			8		5	

					9			6
6	5				8		2	
	7				3	5		
		8	3					4
	3						1	
1					6	3		
		5	7				4	
	1		4				5	9
3			9					

		4	1			7		
6		1		2				
				4		9		
8					2		4	
3			6		4			9
	5		8					2
		9		8				
				7		2		5
		5			3	4		

4					2		8	3
					9	5		
	1							6
					5		6	7
8		5				4		9
3	6		7					
1							5	
		4	3					
6	9		8					1

PUZZLE 255

	5	3		9			8	1
					2			
			7			3		
	2	6		5				
		5	3		4	9		
				7		4	2	
		8			5			
			1					
6	3			8		5	4	

PUZZLE 256

				8				5
						4		
	7	6	4			8	1	
			3					9
	6	4	8		5	1	3	
1					6			
	1	5			4	3	9	
		7						
2				9				

				1		5		2
	1				8		6	
7						3		
				3		9		8
	7		1		9		5	
3		8		2				
		3						6
	6		8				9	
8		1		9				

1		3	8					
				2				9
	2			7		3		
	7	9	2					3
		4				5		
3					1	4	8	
		6		1			3	
4				9				
					5	2		4

3			6			7		
					1		5	6
		7						4
			9	8		6		
4		5				2		7
	1		7	5				
9						4		
5	8		9					
		3			7			1

5			6	3				
	7			9				1
	1			5		2		
				3	6	5		
3								9
	5	9	1					
	8		5				3	
2			7				6	
				2	6			8

9		4				8		
	2							
				9	7	4		2
8		6		3			4	5
2	5			1		7		3
7		5	6	4				
							7	
		8				2		1

5		4					1	
	9		3	1				4
8		1	7					
		5			8			
6								3
			2			9		
					3	1		5
9				7	5		4	
	5					2		7

				4		5		7
7			6					
					1	3	4	
6			8			7		
	9	2				4	8	
		1		3				9
	8	6	2					
				6				1
2		7	5					

				7				
		3		6				8
		9	8		1		7	
3							6	5
	6	2				9	8	
4	9							7
	7		5		9	1		
2			4			8		
			6					

	4			5	7			
	8	5						
6			4		1	7		
	2	4						1
			1		8			
9						6	5	
		1	6		2			9
						4	6	
			5	3			1	

				5				3
						1	9	
4	6				3	5		
3				2	5			
	4						8	
			8	1				7
		4	9				1	8
	9	8						
2				3				

PUZZLE 267

7	8				1			
			6				3	
		5		8		1		
			7			2		3
	3		2		6		7	
1		2			4			
		3		9		8		
	5				8			
			5				4	6

PUZZLE 268

		2			4	9	8	
	7			6				2
9			8	5		2		
	5		4		1		6	
		1		2	6			8
5				1			4	
	4	6	7			5		

2		1		9			7	5
					6			
7						2		8
9					7		1	3
5	8		6					2
8		4						9
			8					
3	9			5		8		4

				3			4	
		1	4				9	
		4			8		2	7
				9		2		
7	6						3	9
		8		2				
3	4		1			5		
	2				6	3		
	1			5				

2				6	3		7	
		4						
	8		4	7		5		
5	6		9					
		7				8		
					1		5	6
		2		4	7		8	
						7		
	9		5	8				3

	4				7	6		5
				8	5			
		1						4
		5	9		1		6	
1								9
	7		5		6	4		
4						1		
			6	5				
3		2	8				5	

2	1							
			7					
8			5		1	6		3
	4			3		8		
9	6						2	4
		2		7			3	
3		1	8		9			6
					5			
							8	2

				5		7		
	7		3			9		
6		9	1				3	
	5		8	9				
4								2
				1	2		8	
	1				8	4		9
		8			1		6	
		3		6				

PUZZLE 275

	7					2		6
						7	4	
				1	7	9	8	
					8		7	
2			6		3			4
	6		4					
	8	3	7	5				
	5	1						
7		9					1	

PUZZLE 276

2		7		8				
6					5			
		3		9		2		
		6		5			8	
3			7		2			9
	4			3		7		
		5		1		8		
			8					5
				4		3		6

	1		7					
		5	6	3		4		9
	4			8				
3		4						7
	6						4	
9						6		8
				9			6	
1		2		5	6	3		
					4		1	

		2			7			
	3	6	1				5	
				5		8		
	8							7
	1	4	3		5	9	8	
6							3	
		5		1				
	9				4	3	2	
			8			4		

PUZZLE 279

				4		9		1
2		3						
		9	6		8			
3		6	8					2
4								7
9					7	6		3
			9		2	8		
						2		4
8		5		3				

PUZZLE 280

1			7					
							7	1
8		6			1		4	
			6				3	
5		1		4		9		2
	9				2			
	3		4			7		9
4	6							
					5			4

1								
			7			1		6
	6		8	9			7	
		9		4				5
	8		6		3		9	
5				7		8		
	4			8	9		5	
6		2			5			
								3

3	2		6					
		8			3		6	9
5				8				
6		7			8		9	
	9		2			7		1
				4				6
2	5		8			4		
					6		5	3

PUZZLE 283

					6	1		7
		7	4				6	
8			9					
		4		9			2	
2		3				9		1
	7			2		6		
					9			5
	8				4	3		
4		6	3					

PUZZLE 284

		6	2		3	9		
				7	4			
5								
6		4	5					9
	3			2			7	
7					1	4		8
								5
			8	4				
		3	1		2	6		

			8		6			
9			2	3			1	
		3		7		4		
7				1				
	3	4				9	7	
				9				5
		2		4		6		
	5			8	1			9
			6		7			

6		1		7	2			5
	8				6	9		
					5		3	
							4	1
				3				
2	5							
	9		5					
		3	8				7	
8			6	9		3		2

			8		6			9
	5	3		1		4		
	6				5			
			3	9			4	
		7				6		
	2			6	8			
			2				8	
		9		4		3	5	
5			7		3			

9			4					
4		1		9	8			
		2			6			
	8	9			5			3
	1						7	
7			2			9	4	
			9			1		
			8	1		2		4
					4			6

3		2	5	6			9	
	4							
8		5		7				
	1			5		8		9
6		3		4			7	
				9		6		8
							2	
	6			8	2	7		5

5						1		
6	3		9					7
	9			6	5			
		8		7	9			
			6		2			
			8	4		7		
			7	1			6	
1					6		7	3
		5						4

PUZZLE 291

		6		7	8	1		
2								
			2	6		3		
	7				2			9
8		9				2		5
1			6				4	
		4		5	6			
								7
		1	8	3		9		

PUZZLE 292

2				8		6		
4					9	7		3
5					7		2	
	5							
9		3				4		8
							7	
	9		6					4
3		6	1					2
		4		5				7

			3				5	
			4	2	9			3
		6						9
		9			8		6	
	6	4				3	7	
	7		1			5		
3						9		
4			5	1	7			
	5				3			

	6						7	
					1	8		
4		7					2	5
9				7			1	
6				3				2
	5			8				6
5	9					2		3
		2	8					
	3						5	

PUZZLE 295

				9			6	
	2	6	5					
1					7	2	5	
	4	8						
		1	9		6	8		
						9	3	
	6	3	4					2
					2	4	8	
	5			1				

PUZZLE 296

	1			2		8		
5	8	9			6			
7							3	9
	4				1			
		6				3		
			3				5	
9	2							3
			9			4	7	1
		1		8			2	

2			5	6				
		7			1	3		6
	1					8		
			2					
9		8	1		3	6		5
				4				
		4					9	
5		1	6			7		
			4	5				3

		8		6	1	5		
			7		5	8		
2								7
5			3					8
	3						9	
9					7			5
8								2
		6	4		9			
		2	1	8		9		

7								
						5		6
	5		3		7	2		
9		1			2	6		
3			9		4			2
		4	5			3		7
		2	1		9		4	
1		3						
								5

	5		7			6		
1								4
		4			1			
9		6		5			4	
		3	6		9	8		
	7			8		3		6
			9			7		
5								1
		1			2		3	

SU
DO
KU

SOLUTION

SOLUTION 1

1	2	7	4	8	6	3	5	9
5	3	8	7	2	9	4	1	6
4	9	6	3	1	5	8	7	2
2	5	3	1	7	8	9	6	4
6	7	1	2	9	4	5	8	3
8	4	9	5	6	3	7	2	1
7	8	4	6	3	1	2	9	5
9	1	5	8	4	2	6	3	7
3	6	2	9	5	7	1	4	8

SOLUTION 2

5	6	4	3	8	9	7	1	2
7	2	9	6	5	1	3	4	8
8	1	3	7	2	4	9	5	6
1	3	8	9	6	7	5	2	4
4	9	7	5	1	2	6	8	3
6	5	2	8	4	3	1	9	7
3	8	1	2	9	6	4	7	5
2	4	6	1	7	5	8	3	9
9	7	5	4	3	8	2	6	1

SOLUTION 3

6	2	5	9	1	7	4	8	3
3	1	4	5	2	8	7	6	9
7	8	9	6	4	3	2	5	1
9	7	1	4	8	6	5	3	2
4	3	2	1	9	5	6	7	8
8	5	6	3	7	2	9	1	4
1	6	7	2	3	4	8	9	5
5	4	3	8	6	9	1	2	7
2	9	8	7	5	1	3	4	6

SOLUTION 4

3	1	7	8	6	5	4	2	9
4	6	9	3	1	2	5	7	8
5	8	2	4	7	9	1	6	3
1	3	4	7	8	6	2	9	5
6	7	5	2	9	3	8	4	1
2	9	8	1	5	4	7	3	6
9	5	1	6	2	7	3	8	4
7	4	6	5	3	8	9	1	2
8	2	3	9	4	1	6	5	7

SOLUTION 5

3	9	5	4	2	6	1	8	7
2	7	8	1	9	3	4	5	6
1	4	6	7	8	5	3	9	2
4	2	7	3	1	8	9	6	5
9	6	1	2	5	4	7	3	8
5	8	3	6	7	9	2	4	1
7	3	9	8	6	2	5	1	4
6	1	4	5	3	7	8	2	9
8	5	2	9	4	1	6	7	3

SOLUTION 6

9	3	6	5	7	8	2	4	1
2	4	7	1	3	9	5	6	8
8	5	1	4	2	6	3	7	9
7	9	4	3	5	1	8	2	6
6	2	8	9	4	7	1	5	3
3	1	5	8	6	2	4	9	7
1	6	2	7	8	5	9	3	4
5	8	3	6	9	4	7	1	2
4	7	9	2	1	3	6	8	5

SOLUTION 7

8	2	4	1	3	7	9	6	5
1	6	3	2	5	9	7	4	8
5	7	9	8	4	6	2	3	1
4	3	5	6	2	8	1	7	9
7	8	1	3	9	4	6	5	2
2	9	6	7	1	5	3	8	4
6	4	2	5	7	1	8	9	3
3	5	8	9	6	2	4	1	7
9	1	7	4	8	3	5	2	6

SOLUTION 8

7	9	5	8	6	3	2	1	4
3	8	2	4	7	1	6	5	9
6	4	1	2	5	9	3	8	7
8	1	3	9	2	4	7	6	5
4	5	9	7	1	6	8	3	2
2	7	6	3	8	5	9	4	1
5	3	8	1	9	7	4	2	6
1	2	7	6	4	8	5	9	3
9	6	4	5	3	2	1	7	8

SOLUTION 9

2	9	8	5	4	3	7	1	6
6	3	1	7	9	8	5	4	2
7	4	5	6	1	2	8	3	9
4	1	3	9	5	6	2	7	8
9	8	7	1	2	4	3	6	5
5	6	2	3	8	7	4	9	1
3	2	6	8	7	1	9	5	4
8	7	9	4	6	5	1	2	3
1	5	4	2	3	9	6	8	7

SOLUTION 10

1	2	7	9	4	5	3	6	8
5	4	8	3	6	1	7	2	9
6	9	3	8	7	2	1	4	5
2	1	4	5	3	8	6	9	7
9	7	5	6	2	4	8	3	1
3	8	6	1	9	7	2	5	4
4	3	1	2	8	9	5	7	6
8	6	9	7	5	3	4	1	2
7	5	2	4	1	6	9	8	3

SOLUTION 11

3	2	6	7	4	5	1	8	9
4	5	1	8	6	9	7	3	2
7	8	9	1	2	3	6	4	5
6	7	4	3	1	2	5	9	8
1	9	5	4	7	8	2	6	3
2	3	8	5	9	6	4	1	7
9	4	2	6	3	7	8	5	1
5	6	3	2	8	1	9	7	4
8	1	7	9	5	4	3	2	6

SOLUTION 12

8	2	9	1	4	3	6	7	5
6	5	3	7	9	8	2	4	1
7	1	4	2	5	6	3	8	9
1	3	8	9	7	2	4	5	6
2	9	5	4	6	1	7	3	8
4	7	6	3	8	5	9	1	2
3	4	2	5	1	9	8	6	7
5	6	7	8	2	4	1	9	3
9	8	1	6	3	7	5	2	4

SOLUTION 13

4	2	9	7	3	6	5	8	1
1	6	7	4	8	5	2	9	3
3	8	5	1	9	2	6	7	4
2	5	3	6	4	8	9	1	7
6	1	8	5	7	9	4	3	2
9	7	4	3	2	1	8	6	5
7	3	2	9	6	4	1	5	8
5	4	6	8	1	3	7	2	9
8	9	1	2	5	7	3	4	6

SOLUTION 14

2	7	5	8	1	3	4	6	9
1	3	4	6	2	9	5	7	8
6	8	9	4	5	7	3	1	2
3	2	7	5	8	1	6	9	4
5	9	6	3	7	4	2	8	1
8	4	1	2	9	6	7	5	3
4	5	3	1	6	8	9	2	7
9	6	8	7	4	2	1	3	5
7	1	2	9	3	5	8	4	6

SOLUTION 15

2	7	6	5	4	9	1	3	8
8	9	3	6	2	1	7	5	4
4	5	1	8	7	3	6	9	2
3	6	5	1	8	7	2	4	9
9	2	7	3	5	4	8	6	1
1	4	8	9	6	2	3	7	5
6	1	2	7	9	5	4	8	3
5	8	4	2	3	6	9	1	7
7	3	9	4	1	8	5	2	6

SOLUTION 16

1	8	2	9	4	5	3	7	6
3	5	9	2	6	7	4	1	8
6	4	7	8	3	1	5	2	9
5	7	8	6	1	9	2	3	4
9	3	6	4	7	2	8	5	1
4	2	1	5	8	3	9	6	7
8	6	3	1	5	4	7	9	2
2	1	5	7	9	8	6	4	3
7	9	4	3	2	6	1	8	5

SOLUTION 17

8	4	1	2	3	5	7	9	6
7	3	2	9	6	4	1	8	5
5	9	6	8	7	1	4	2	3
2	6	4	7	5	3	9	1	8
1	7	9	4	8	6	3	5	2
3	5	8	1	2	9	6	4	7
4	8	3	5	1	7	2	6	9
9	2	7	6	4	8	5	3	1
6	1	5	3	9	2	8	7	4

SOLUTION 18

9	2	7	3	4	5	8	1	6
3	8	1	7	2	6	9	5	4
4	6	5	1	9	8	3	7	2
5	1	8	2	7	4	6	3	9
7	3	9	8	6	1	4	2	5
2	4	6	5	3	9	1	8	7
1	5	4	9	8	7	2	6	3
8	9	2	6	5	3	7	4	1
6	7	3	4	1	2	5	9	8

SOLUTION 19

5	6	2	9	7	4	8	1	3
4	7	1	3	5	8	2	9	6
8	9	3	2	1	6	5	4	7
3	2	4	6	8	1	9	7	5
6	5	8	7	3	9	1	2	4
7	1	9	4	2	5	3	6	8
2	4	6	8	9	3	7	5	1
1	8	7	5	4	2	6	3	9
9	3	5	1	6	7	4	8	2

SOLUTION 20

4	2	1	7	8	6	3	5	9
5	7	8	2	3	9	4	1	6
9	3	6	1	4	5	7	2	8
6	5	7	8	1	3	9	4	2
3	9	2	5	7	4	8	6	1
1	8	4	6	9	2	5	3	7
8	4	9	3	2	1	6	7	5
7	1	5	4	6	8	2	9	3
2	6	3	9	5	7	1	8	4

SOLUTION 21

6	3	2	9	1	8	4	7	5
7	8	5	2	3	4	1	6	9
1	4	9	7	5	6	3	2	8
4	5	6	8	9	7	2	1	3
9	2	7	1	4	3	5	8	6
8	1	3	5	6	2	9	4	7
3	7	1	4	8	9	6	5	2
5	6	8	3	2	1	7	9	4
2	9	4	6	7	5	8	3	1

SOLUTION 22

7	1	5	2	9	6	3	4	8
9	8	2	3	7	4	6	5	1
3	4	6	1	5	8	7	2	9
2	7	9	5	3	1	8	6	4
4	6	1	9	8	2	5	3	7
5	3	8	6	4	7	9	1	2
6	9	7	4	1	3	2	8	5
8	2	4	7	6	5	1	9	3
1	5	3	8	2	9	4	7	6

SOLUTION 23

9	1	2	7	8	4	3	5	6
5	6	8	3	1	2	4	9	7
3	4	7	9	5	6	1	2	8
8	3	4	5	9	7	2	6	1
7	9	1	2	6	8	5	3	4
6	2	5	4	3	1	7	8	9
2	5	6	1	7	9	8	4	3
1	8	3	6	4	5	9	7	2
4	7	9	8	2	3	6	1	5

SOLUTION 24

6	9	5	2	7	8	1	4	3
2	4	3	9	1	5	8	7	6
8	7	1	6	3	4	2	9	5
7	2	6	3	4	1	9	5	8
5	3	4	7	8	9	6	2	1
9	1	8	5	6	2	7	3	4
3	8	7	4	9	6	5	1	2
1	5	9	8	2	3	4	6	7
4	6	2	1	5	7	3	8	9

SOLUTION 25

5	7	8	4	2	3	1	9	6
3	2	1	8	6	9	5	4	7
6	4	9	7	5	1	3	2	8
1	8	4	6	3	5	2	7	9
2	9	3	1	4	7	8	6	5
7	5	6	2	9	8	4	3	1
4	3	7	5	1	6	9	8	2
9	6	5	3	8	2	7	1	4
8	1	2	9	7	4	6	5	3

SOLUTION 26

9	1	2	5	7	3	6	4	8
5	7	8	6	1	4	2	3	9
3	6	4	9	8	2	1	5	7
7	3	6	1	5	9	8	2	4
2	9	5	8	4	7	3	6	1
4	8	1	3	2	6	7	9	5
1	5	3	4	6	8	9	7	2
8	2	9	7	3	5	4	1	6
6	4	7	2	9	1	5	8	3

SOLUTION 27

1	4	8	5	6	2	9	3	7
9	7	3	1	8	4	2	6	5
2	6	5	7	9	3	4	1	8
5	8	4	9	7	6	1	2	3
7	2	9	3	5	1	8	4	6
6	3	1	4	2	8	7	5	9
4	1	7	6	3	9	5	8	2
8	9	6	2	4	5	3	7	1
3	5	2	8	1	7	6	9	4

SOLUTION 28

7	2	6	9	4	3	8	5	1
4	1	3	6	5	8	2	9	7
5	8	9	1	7	2	3	6	4
6	3	2	4	1	5	7	8	9
8	7	4	3	9	6	1	2	5
1	9	5	8	2	7	4	3	6
2	5	1	7	3	9	6	4	8
3	4	8	5	6	1	9	7	2
9	6	7	2	8	4	5	1	3

SOLUTION 29

2	6	5	9	4	1	8	7	3
3	4	1	5	7	8	6	2	9
9	8	7	6	3	2	5	1	4
1	5	4	2	8	3	7	9	6
7	2	9	1	6	4	3	5	8
6	3	8	7	9	5	1	4	2
4	1	6	3	5	9	2	8	7
8	7	2	4	1	6	9	3	5
5	9	3	8	2	7	4	6	1

SOLUTION 30

2	4	1	8	3	9	6	7	5
3	5	6	2	1	7	9	8	4
7	9	8	4	5	6	2	1	3
8	3	4	9	6	5	7	2	1
1	6	5	7	2	3	4	9	8
9	7	2	1	4	8	5	3	6
6	1	9	3	7	4	8	5	2
4	8	3	5	9	2	1	6	7
5	2	7	6	8	1	3	4	9

SOLUTION 31

3	7	1	8	9	4	2	5	6
5	2	6	7	1	3	4	9	8
4	8	9	5	2	6	7	3	1
7	4	5	3	6	1	9	8	2
2	6	3	4	8	9	5	1	7
1	9	8	2	5	7	6	4	3
8	3	7	9	4	2	1	6	5
6	5	4	1	7	8	3	2	9
9	1	2	6	3	5	8	7	4

SOLUTION 32

5	6	1	2	7	4	3	9	8
7	3	4	1	8	9	2	5	6
9	2	8	6	5	3	7	4	1
1	8	3	7	4	6	5	2	9
6	5	2	3	9	8	4	1	7
4	7	9	5	2	1	6	8	3
3	9	7	4	1	2	8	6	5
2	1	6	8	3	5	9	7	4
8	4	5	9	6	7	1	3	2

SOLUTION 33

1	6	2	4	8	7	3	9	5
7	9	4	6	5	3	8	1	2
8	5	3	9	2	1	4	7	6
5	2	6	7	1	4	9	3	8
4	8	1	2	3	9	5	6	7
9	3	7	8	6	5	1	2	4
6	4	9	1	7	8	2	5	3
3	7	8	5	9	2	6	4	1
2	1	5	3	4	6	7	8	9

SOLUTION 34

6	4	2	3	7	1	8	9	5
8	5	3	2	9	4	6	1	7
1	9	7	8	5	6	2	3	4
2	3	5	1	4	8	9	7	6
9	8	4	5	6	7	3	2	1
7	6	1	9	2	3	5	4	8
3	1	9	4	8	5	7	6	2
5	2	6	7	1	9	4	8	3
4	7	8	6	3	2	1	5	9

SOLUTION 35

9	5	8	3	1	7	4	6	2
2	3	7	6	4	5	1	9	8
1	4	6	8	9	2	7	3	5
3	9	5	1	8	4	6	2	7
4	7	1	2	6	9	5	8	3
6	8	2	5	7	3	9	4	1
5	1	3	4	2	6	8	7	9
7	2	4	9	5	8	3	1	6
8	6	9	7	3	1	2	5	4

SOLUTION 36

5	2	6	1	3	9	8	4	7
4	1	3	2	8	7	5	9	6
8	9	7	5	4	6	3	1	2
6	3	8	9	2	4	1	7	5
1	5	9	6	7	3	2	8	4
2	7	4	8	5	1	6	3	9
7	4	1	3	6	5	9	2	8
9	8	5	7	1	2	4	6	3
3	6	2	4	9	8	7	5	1

SOLUTION 37

4	8	1	6	9	7	2	3	5
6	5	9	3	4	2	7	8	1
2	7	3	1	5	8	9	4	6
3	6	5	8	1	9	4	2	7
9	1	2	5	7	4	8	6	3
8	4	7	2	3	6	5	1	9
1	2	6	7	8	5	3	9	4
7	3	4	9	2	1	6	5	8
5	9	8	4	6	3	1	7	2

SOLUTION 38

8	1	2	3	9	4	5	6	7
4	6	7	2	8	5	1	9	3
9	3	5	1	6	7	8	2	4
1	8	4	9	7	2	3	5	6
3	2	9	4	5	6	7	1	8
7	5	6	8	3	1	9	4	2
6	9	3	5	4	8	2	7	1
5	7	1	6	2	3	4	8	9
2	4	8	7	1	9	6	3	5

SOLUTION 39

8	3	6	2	1	7	5	4	9
4	9	2	8	5	6	1	7	3
5	1	7	4	9	3	2	6	8
7	8	9	3	4	5	6	1	2
6	5	1	7	8	2	9	3	4
3	2	4	1	6	9	7	8	5
2	6	3	5	7	8	4	9	1
1	7	5	9	3	4	8	2	6
9	4	8	6	2	1	3	5	7

SOLUTION 40

2	5	3	7	4	6	8	1	9
6	9	8	1	3	2	7	4	5
7	1	4	5	9	8	3	2	6
1	3	2	4	5	7	6	9	8
8	7	9	3	6	1	2	5	4
5	4	6	2	8	9	1	3	7
4	2	7	6	1	5	9	8	3
3	8	1	9	7	4	5	6	2
9	6	5	8	2	3	4	7	1

SOLUTION 41

5	2	3	6	1	8	4	7	9
4	1	6	9	3	7	2	5	8
9	8	7	4	2	5	6	3	1
2	6	8	3	4	1	5	9	7
3	5	9	7	6	2	1	8	4
1	7	4	8	5	9	3	6	2
7	4	5	1	9	3	8	2	6
8	3	1	2	7	6	9	4	5
6	9	2	5	8	4	7	1	3

SOLUTION 42

8	7	4	2	1	9	3	6	5
5	6	2	8	3	7	1	4	9
1	9	3	4	5	6	7	2	8
2	1	9	3	6	8	5	7	4
3	4	8	7	9	5	6	1	2
7	5	6	1	4	2	8	9	3
9	2	7	6	8	3	4	5	1
6	8	1	5	2	4	9	3	7
4	3	5	9	7	1	2	8	6

SOLUTION 43

2	8	5	3	7	4	9	1	6
9	6	3	2	8	1	4	7	5
4	1	7	6	5	9	3	8	2
7	3	9	1	4	5	2	6	8
1	4	8	7	6	2	5	9	3
5	2	6	8	9	3	7	4	1
3	9	2	4	1	6	8	5	7
8	5	1	9	2	7	6	3	4
6	7	4	5	3	8	1	2	9

SOLUTION 44

5	1	8	6	4	7	9	2	3
7	4	2	8	3	9	6	1	5
9	6	3	1	2	5	7	4	8
4	7	5	2	6	8	1	3	9
1	8	6	5	9	3	2	7	4
3	2	9	4	7	1	5	8	6
6	5	4	7	8	2	3	9	1
2	3	1	9	5	4	8	6	7
8	9	7	3	1	6	4	5	2

SOLUTION 45

5	1	2	4	7	6	3	9	8
6	7	9	1	3	8	2	4	5
4	8	3	9	2	5	6	7	1
3	5	6	8	4	9	7	1	2
9	4	1	2	5	7	8	6	3
8	2	7	3	6	1	9	5	4
7	6	8	5	1	2	4	3	9
2	3	5	6	9	4	1	8	7
1	9	4	7	8	3	5	2	6

SOLUTION 46

9	3	4	7	2	5	8	1	6
1	6	7	8	3	9	2	4	5
8	5	2	1	6	4	7	3	9
4	2	1	9	7	6	3	5	8
7	9	3	5	8	1	6	2	4
5	8	6	2	4	3	9	7	1
3	4	8	6	1	7	5	9	2
6	1	5	3	9	2	4	8	7
2	7	9	4	5	8	1	6	3

SOLUTION 47

1	7	3	6	9	4	2	8	5
2	4	8	5	3	1	9	6	7
9	6	5	7	8	2	1	3	4
7	1	6	8	4	3	5	2	9
3	5	4	2	6	9	8	7	1
8	9	2	1	7	5	3	4	6
4	2	1	3	5	6	7	9	8
6	3	7	9	1	8	4	5	2
5	8	9	4	2	7	6	1	3

SOLUTION 48

1	6	9	4	3	8	2	7	5
7	4	3	6	2	5	8	9	1
5	8	2	7	9	1	6	3	4
3	5	8	1	6	7	4	2	9
2	1	7	9	5	4	3	8	6
4	9	6	2	8	3	1	5	7
8	3	4	5	7	6	9	1	2
9	7	1	8	4	2	5	6	3
6	2	5	3	1	9	7	4	8

SOLUTION 49

2	1	3	4	5	6	7	9	8
8	9	7	2	3	1	4	5	6
5	4	6	9	8	7	2	3	1
9	5	1	3	2	8	6	4	7
3	7	2	6	9	4	1	8	5
6	8	4	7	1	5	3	2	9
4	3	8	1	6	9	5	7	2
7	6	5	8	4	2	9	1	3
1	2	9	5	7	3	8	6	4

SOLUTION 50

5	3	1	6	2	8	7	4	9
7	4	6	9	3	5	2	1	8
2	8	9	1	4	7	6	3	5
9	1	4	3	6	2	8	5	7
8	5	3	7	1	9	4	2	6
6	2	7	5	8	4	1	9	3
3	6	2	8	5	1	9	7	4
1	9	5	4	7	6	3	8	2
4	7	8	2	9	3	5	6	1

SOLUTION 51

7	2	9	6	4	8	3	1	5
5	8	6	1	2	3	9	4	7
3	4	1	7	9	5	2	8	6
2	7	3	9	6	4	1	5	8
1	9	4	8	5	7	6	3	2
6	5	8	3	1	2	7	9	4
4	1	2	5	7	9	8	6	3
8	6	5	2	3	1	4	7	9
9	3	7	4	8	6	5	2	1

SOLUTION 52

4	9	7	2	8	3	6	1	5
2	5	3	4	1	6	7	8	9
1	8	6	7	5	9	4	2	3
7	4	1	6	2	5	9	3	8
6	2	5	3	9	8	1	4	7
9	3	8	1	4	7	5	6	2
8	6	4	9	7	2	3	5	1
3	7	2	5	6	1	8	9	4
5	1	9	8	3	4	2	7	6

SOLUTION 53

7	9	8	1	6	3	2	5	4
6	4	1	5	2	7	3	9	8
2	3	5	4	8	9	7	6	1
3	7	2	6	5	4	8	1	9
9	8	6	2	3	1	4	7	5
5	1	4	9	7	8	6	2	3
8	5	9	7	4	6	1	3	2
1	6	3	8	9	2	5	4	7
4	2	7	3	1	5	9	8	6

SOLUTION 54

3	6	8	5	2	1	4	9	7
5	4	1	3	9	7	2	6	8
2	7	9	8	4	6	1	3	5
4	1	3	2	8	5	6	7	9
7	8	6	1	3	9	5	2	4
9	5	2	7	6	4	8	1	3
1	2	4	9	7	8	3	5	6
6	3	7	4	5	2	9	8	1
8	9	5	6	1	3	7	4	2

SOLUTION 55

6	1	3	4	8	2	5	9	7
5	4	7	1	3	9	6	2	8
8	2	9	7	5	6	4	1	3
4	8	1	5	9	3	2	7	6
9	5	6	8	2	7	1	3	4
7	3	2	6	1	4	9	8	5
3	7	5	9	4	1	8	6	2
1	6	8	2	7	5	3	4	9
2	9	4	3	6	8	7	5	1

SOLUTION 56

8	1	4	6	3	5	2	7	9
7	9	5	2	4	8	3	1	6
2	3	6	7	9	1	4	5	8
4	8	1	5	7	6	9	3	2
5	2	3	8	1	9	7	6	4
9	6	7	3	2	4	5	8	1
3	4	9	1	6	7	8	2	5
1	5	2	4	8	3	6	9	7
6	7	8	9	5	2	1	4	3

SOLUTION 57

1	8	3	6	2	9	5	4	7
9	5	4	8	1	7	3	6	2
2	7	6	4	3	5	9	8	1
3	6	5	2	7	4	8	1	9
7	2	1	3	9	8	4	5	6
4	9	8	1	5	6	2	7	3
8	3	2	7	4	1	6	9	5
5	4	7	9	6	3	1	2	8
6	1	9	5	8	2	7	3	4

SOLUTION 58

7	2	8	4	1	9	3	5	6
4	6	1	5	7	3	2	8	9
9	5	3	6	8	2	1	4	7
3	8	6	7	5	1	4	9	2
2	1	4	3	9	8	7	6	5
5	9	7	2	6	4	8	3	1
1	3	2	9	4	5	6	7	8
6	4	9	8	2	7	5	1	3
8	7	5	1	3	6	9	2	4

SOLUTION 59

2	7	9	6	5	3	4	1	8
3	4	5	1	9	8	2	6	7
1	6	8	4	2	7	3	5	9
4	8	2	9	6	1	5	7	3
7	5	6	3	4	2	9	8	1
9	1	3	7	8	5	6	4	2
5	9	7	8	3	4	1	2	6
8	3	4	2	1	6	7	9	5
6	2	1	5	7	9	8	3	4

SOLUTION 60

2	8	9	6	3	1	7	5	4
6	1	4	7	8	5	2	3	9
5	3	7	4	9	2	8	6	1
7	4	5	2	6	9	3	1	8
8	6	1	5	7	3	9	4	2
9	2	3	1	4	8	5	7	6
4	7	2	9	5	6	1	8	3
3	9	6	8	1	7	4	2	5
1	5	8	3	2	4	6	9	7

SOLUTION 61

9	6	3	8	5	7	2	1	4
1	8	5	9	4	2	6	3	7
7	2	4	3	6	1	9	8	5
4	9	6	7	3	5	1	2	8
5	1	7	2	8	6	4	9	3
2	3	8	1	9	4	5	7	6
3	4	2	5	1	8	7	6	9
8	5	1	6	7	9	3	4	2
6	7	9	4	2	3	8	5	1

SOLUTION 62

6	3	7	4	9	5	2	8	1
4	1	5	2	3	8	7	6	9
2	8	9	6	7	1	3	4	5
1	5	2	9	6	3	4	7	8
3	6	8	1	4	7	5	9	2
7	9	4	5	8	2	1	3	6
8	4	1	3	5	6	9	2	7
9	2	6	7	1	4	8	5	3
5	7	3	8	2	9	6	1	4

SOLUTION 63

2	7	1	5	8	3	9	6	4
8	3	9	7	6	4	2	1	5
5	6	4	1	2	9	8	3	7
7	9	2	8	5	6	1	4	3
4	5	3	2	9	1	6	7	8
6	1	8	3	4	7	5	2	9
1	8	7	6	3	5	4	9	2
3	4	5	9	1	2	7	8	6
9	2	6	4	7	8	3	5	1

SOLUTION 64

8	4	3	7	6	2	1	5	9
9	5	1	4	8	3	6	2	7
7	2	6	1	9	5	8	3	4
2	8	7	3	4	9	5	6	1
3	9	4	5	1	6	7	8	2
6	1	5	8	2	7	4	9	3
5	6	9	2	7	4	3	1	8
4	3	8	9	5	1	2	7	6
1	7	2	6	3	8	9	4	5

SOLUTION 65

3	6	1	2	4	5	7	8	9
2	4	9	7	8	3	1	5	6
8	5	7	1	6	9	2	4	3
6	2	4	3	7	1	8	9	5
5	7	3	4	9	8	6	2	1
9	1	8	5	2	6	4	3	7
4	8	6	9	5	7	3	1	2
7	3	5	8	1	2	9	6	4
1	9	2	6	3	4	5	7	8

SOLUTION 66

6	2	5	1	7	3	8	9	4
8	4	3	9	2	5	1	6	7
9	1	7	4	8	6	3	5	2
4	5	1	8	3	2	9	7	6
3	9	8	7	6	4	5	2	1
7	6	2	5	1	9	4	8	3
2	3	9	6	4	8	7	1	5
5	7	6	3	9	1	2	4	8
1	8	4	2	5	7	6	3	9

SOLUTION 67

8	7	6	2	4	1	9	3	5
2	4	5	3	8	9	6	1	7
3	1	9	7	6	5	8	2	4
9	2	4	8	5	7	1	6	3
7	3	8	1	9	6	4	5	2
6	5	1	4	3	2	7	9	8
1	6	7	5	2	4	3	8	9
4	8	2	9	1	3	5	7	6
5	9	3	6	7	8	2	4	1

SOLUTION 68

5	1	4	2	7	9	8	3	6
8	6	9	5	3	4	1	2	7
3	7	2	6	1	8	5	9	4
9	5	8	7	4	2	6	1	3
4	3	7	1	8	6	9	5	2
6	2	1	3	9	5	4	7	8
1	9	3	4	6	7	2	8	5
7	4	5	8	2	1	3	6	9
2	8	6	9	5	3	7	4	1

SOLUTION 69

8	7	9	3	4	6	2	5	1
5	6	3	7	1	2	8	9	4
2	1	4	8	9	5	6	3	7
3	5	8	1	6	4	7	2	9
9	4	1	5	2	7	3	8	6
6	2	7	9	3	8	1	4	5
1	8	5	4	7	3	9	6	2
7	3	6	2	5	9	4	1	8
4	9	2	6	8	1	5	7	3

SOLUTION 70

3	2	4	7	1	8	9	6	5
5	1	7	2	6	9	4	3	8
8	6	9	3	5	4	7	2	1
7	8	1	5	4	6	2	9	3
4	3	5	9	7	2	8	1	6
2	9	6	8	3	1	5	4	7
1	5	8	4	2	3	6	7	9
6	7	2	1	9	5	3	8	4
9	4	3	6	8	7	1	5	2

SOLUTION 71

1	4	6	9	3	8	2	7	5
9	2	8	6	7	5	3	1	4
5	7	3	2	4	1	6	8	9
6	3	1	4	8	7	9	5	2
2	9	7	1	5	6	4	3	8
8	5	4	3	9	2	7	6	1
3	8	5	7	2	4	1	9	6
7	6	2	8	1	9	5	4	3
4	1	9	5	6	3	8	2	7

SOLUTION 72

4	7	5	1	2	9	6	3	8
3	8	1	6	5	7	4	2	9
6	9	2	3	4	8	1	7	5
5	2	7	9	6	1	8	4	3
1	6	8	2	3	4	9	5	7
9	4	3	7	8	5	2	1	6
8	1	9	4	7	3	5	6	2
7	5	6	8	1	2	3	9	4
2	3	4	5	9	6	7	8	1

SOLUTION 73

7	3	8	1	9	6	2	5	4
4	5	2	3	8	7	6	9	1
9	1	6	4	2	5	3	7	8
1	6	9	2	4	3	5	8	7
3	7	5	9	6	8	4	1	2
2	8	4	5	7	1	9	3	6
6	9	3	7	1	4	8	2	5
5	4	1	8	3	2	7	6	9
8	2	7	6	5	9	1	4	3

SOLUTION 74

8	6	4	7	2	5	3	9	1
2	5	7	3	1	9	4	8	6
1	3	9	4	6	8	5	2	7
5	2	8	9	7	4	1	6	3
3	9	1	5	8	6	2	7	4
4	7	6	2	3	1	9	5	8
7	1	5	8	9	3	6	4	2
6	4	2	1	5	7	8	3	9
9	8	3	6	4	2	7	1	5

SOLUTION 75

8	3	1	9	5	7	2	4	6
2	5	7	1	4	6	8	3	9
6	9	4	2	8	3	7	1	5
4	1	5	3	2	8	6	9	7
7	6	9	4	1	5	3	8	2
3	2	8	7	6	9	1	5	4
9	4	6	8	7	1	5	2	3
1	7	3	5	9	2	4	6	8
5	8	2	6	3	4	9	7	1

SOLUTION 76

1	6	4	2	5	3	9	8	7
3	8	2	4	7	9	6	1	5
9	7	5	8	1	6	3	2	4
6	5	1	3	8	2	7	4	9
7	2	3	9	4	1	8	5	6
4	9	8	5	6	7	1	3	2
8	3	9	6	2	4	5	7	1
2	1	6	7	3	5	4	9	8
5	4	7	1	9	8	2	6	3

SOLUTION 77

2	6	7	3	1	4	5	9	8
8	3	9	2	6	5	1	4	7
4	1	5	8	9	7	2	3	6
3	5	1	4	7	8	9	6	2
6	9	4	1	2	3	8	7	5
7	2	8	9	5	6	3	1	4
9	4	6	5	8	1	7	2	3
1	8	3	7	4	2	6	5	9
5	7	2	6	3	9	4	8	1

SOLUTION 78

4	2	5	7	6	1	9	3	8
1	6	7	8	9	3	4	2	5
9	8	3	2	4	5	7	1	6
3	4	8	5	7	9	2	6	1
7	9	6	1	8	2	5	4	3
5	1	2	6	3	4	8	7	9
2	7	1	3	5	8	6	9	4
6	5	9	4	1	7	3	8	2
8	3	4	9	2	6	1	5	7

SOLUTION 79

2	7	1	6	3	9	8	5	4
6	3	4	5	7	8	1	9	2
5	8	9	4	2	1	6	7	3
7	1	2	3	9	5	4	6	8
8	4	3	7	6	2	9	1	5
9	6	5	1	8	4	2	3	7
3	9	8	2	1	7	5	4	6
1	5	7	8	4	6	3	2	9
4	2	6	9	5	3	7	8	1

SOLUTION 80

7	9	1	4	5	8	6	2	3
3	5	6	1	2	9	8	4	7
2	4	8	6	3	7	1	5	9
8	6	3	2	4	5	7	9	1
5	2	7	3	9	1	4	8	6
9	1	4	8	7	6	2	3	5
1	3	2	9	6	4	5	7	8
4	8	5	7	1	3	9	6	2
6	7	9	5	8	2	3	1	4

SOLUTION 81

3	1	4	7	6	5	9	2	8
8	2	9	1	3	4	6	5	7
7	5	6	9	8	2	3	1	4
9	8	7	6	1	3	2	4	5
1	3	5	2	4	8	7	9	6
6	4	2	5	7	9	1	8	3
5	9	3	4	2	6	8	7	1
4	6	1	8	9	7	5	3	2
2	7	8	3	5	1	4	6	9

SOLUTION 82

2	6	8	3	5	9	4	1	7
1	5	4	2	6	7	9	3	8
3	9	7	4	8	1	6	2	5
8	3	1	7	2	4	5	9	6
9	4	5	6	1	3	8	7	2
6	7	2	5	9	8	3	4	1
5	8	3	1	4	2	7	6	9
7	1	6	9	3	5	2	8	4
4	2	9	8	7	6	1	5	3

SOLUTION 83

3	8	4	7	2	9	6	1	5
6	7	9	8	1	5	2	3	4
1	2	5	3	4	6	8	7	9
2	3	7	9	5	1	4	8	6
8	4	6	2	7	3	5	9	1
5	9	1	6	8	4	3	2	7
4	1	3	5	9	8	7	6	2
7	5	8	1	6	2	9	4	3
9	6	2	4	3	7	1	5	8

SOLUTION 84

3	1	7	4	2	9	6	5	8
4	6	8	3	5	1	9	2	7
5	9	2	6	7	8	4	1	3
2	3	6	5	4	7	8	9	1
9	8	4	2	1	3	5	7	6
7	5	1	9	8	6	2	3	4
6	4	9	1	3	5	7	8	2
1	7	5	8	6	2	3	4	9
8	2	3	7	9	4	1	6	5

SOLUTION 85

2	3	6	5	1	9	4	7	8
4	5	9	3	8	7	1	2	6
1	7	8	2	4	6	3	5	9
8	2	3	9	6	1	5	4	7
5	1	4	7	3	8	6	9	2
6	9	7	4	2	5	8	3	1
7	8	2	6	5	4	9	1	3
9	6	5	1	7	3	2	8	4
3	4	1	8	9	2	7	6	5

SOLUTION 86

6	3	5	8	9	4	7	1	2
9	4	1	3	7	2	8	6	5
7	2	8	1	6	5	9	4	3
2	1	7	5	3	6	4	9	8
3	6	9	2	4	8	5	7	1
8	5	4	9	1	7	2	3	6
5	7	3	6	2	9	1	8	4
4	8	6	7	5	1	3	2	9
1	9	2	4	8	3	6	5	7

SOLUTION 87

5	3	7	6	1	4	9	8	2
1	6	9	5	2	8	4	3	7
4	2	8	9	3	7	5	1	6
9	8	5	1	7	3	6	2	4
6	7	1	4	5	2	3	9	8
2	4	3	8	6	9	1	7	5
8	5	4	7	9	1	2	6	3
7	1	2	3	4	6	8	5	9
3	9	6	2	8	5	7	4	1

SOLUTION 88

8	7	3	5	4	2	1	9	6
6	2	9	7	3	1	4	8	5
4	5	1	6	9	8	7	3	2
5	4	2	8	1	6	3	7	9
3	8	6	9	2	7	5	1	4
9	1	7	3	5	4	6	2	8
7	9	8	4	6	3	2	5	1
1	3	4	2	8	5	9	6	7
2	6	5	1	7	9	8	4	3

SOLUTION 89

3	1	4	6	7	5	8	9	2
9	8	7	4	3	2	6	5	1
2	6	5	9	1	8	4	7	3
1	3	9	5	8	7	2	4	6
6	7	2	1	9	4	5	3	8
5	4	8	2	6	3	9	1	7
7	5	3	8	4	6	1	2	9
4	9	6	7	2	1	3	8	5
8	2	1	3	5	9	7	6	4

SOLUTION 90

2	9	4	1	3	6	7	8	5
5	6	7	8	2	4	3	9	1
3	1	8	9	7	5	4	2	6
7	5	9	2	1	8	6	4	3
8	3	1	4	6	7	2	5	9
4	2	6	3	5	9	1	7	8
1	7	3	5	8	2	9	6	4
9	8	2	6	4	3	5	1	7
6	4	5	7	9	1	8	3	2

SOLUTION 91

3	9	4	1	5	6	7	2	8
7	2	5	3	4	8	9	6	1
6	1	8	7	9	2	3	4	5
9	6	2	4	7	5	8	1	3
8	7	3	2	1	9	4	5	6
5	4	1	6	8	3	2	7	9
2	5	6	8	3	4	1	9	7
1	8	9	5	2	7	6	3	4
4	3	7	9	6	1	5	8	2

SOLUTION 92

2	4	3	5	9	8	7	6	1
9	7	6	1	4	2	3	8	5
5	1	8	6	3	7	2	9	4
7	5	4	3	2	9	6	1	8
6	9	1	8	7	5	4	3	2
3	8	2	4	6	1	5	7	9
4	6	5	9	1	3	8	2	7
1	3	7	2	8	4	9	5	6
8	2	9	7	5	6	1	4	3

SOLUTION 93

6	4	3	8	9	2	7	5	1
5	8	9	6	7	1	4	3	2
1	7	2	5	3	4	6	8	9
4	1	6	2	5	7	3	9	8
9	3	5	1	6	8	2	4	7
8	2	7	3	4	9	1	6	5
3	5	8	7	2	6	9	1	4
2	9	1	4	8	3	5	7	6
7	6	4	9	1	5	8	2	3

SOLUTION 94

2	6	9	3	5	7	1	4	8
8	3	4	2	6	1	5	9	7
1	7	5	9	8	4	2	3	6
4	1	3	6	7	9	8	2	5
5	8	7	4	1	2	3	6	9
9	2	6	8	3	5	7	1	4
6	4	1	5	2	8	9	7	3
3	5	2	7	9	6	4	8	1
7	9	8	1	4	3	6	5	2

SOLUTION 95

2	6	8	9	4	1	7	5	3
9	4	7	5	2	3	1	8	6
5	1	3	7	6	8	4	9	2
1	3	5	2	9	4	6	7	8
6	8	9	3	7	5	2	1	4
7	2	4	8	1	6	9	3	5
8	9	1	4	3	2	5	6	7
4	5	6	1	8	7	3	2	9
3	7	2	6	5	9	8	4	1

SOLUTION 96

4	3	6	7	5	2	1	9	8
8	2	5	3	9	1	7	6	4
9	7	1	8	4	6	3	5	2
6	8	3	2	1	4	5	7	9
7	5	9	6	3	8	2	4	1
2	1	4	5	7	9	6	8	3
3	4	7	9	2	5	8	1	6
5	9	8	1	6	3	4	2	7
1	6	2	4	8	7	9	3	5

SOLUTION 97

6	1	4	2	8	3	5	9	7
9	7	5	4	6	1	3	2	8
2	8	3	7	9	5	4	1	6
8	3	6	5	4	9	2	7	1
7	9	2	3	1	6	8	4	5
4	5	1	8	7	2	9	6	3
1	4	9	6	5	8	7	3	2
3	6	8	9	2	7	1	5	4
5	2	7	1	3	4	6	8	9

SOLUTION 98

4	5	9	2	7	6	8	3	1
6	8	2	4	3	1	9	7	5
7	3	1	8	5	9	4	6	2
8	2	3	9	4	5	6	1	7
5	4	6	3	1	7	2	8	9
9	1	7	6	2	8	5	4	3
2	7	4	5	6	3	1	9	8
1	9	5	7	8	4	3	2	6
3	6	8	1	9	2	7	5	4

SOLUTION 99

3	9	2	7	8	1	6	4	5
8	6	1	4	9	5	3	2	7
7	4	5	3	2	6	1	9	8
6	2	3	9	1	8	7	5	4
4	7	8	5	6	3	9	1	2
5	1	9	2	4	7	8	6	3
1	3	7	6	5	4	2	8	9
2	5	6	8	7	9	4	3	1
9	8	4	1	3	2	5	7	6

SOLUTION 100

7	6	3	5	4	8	1	2	9
2	4	5	1	3	9	8	7	6
9	1	8	2	6	7	5	4	3
6	3	4	9	5	2	7	1	8
5	7	9	8	1	4	3	6	2
8	2	1	6	7	3	4	9	5
1	9	6	7	8	5	2	3	4
3	5	2	4	9	1	6	8	7
4	8	7	3	2	6	9	5	1

SOLUTION 101

8	9	3	7	1	2	5	4	6
6	5	7	3	9	4	1	2	8
1	2	4	8	6	5	3	9	7
9	4	8	1	2	6	7	3	5
3	1	5	9	8	7	4	6	2
2	7	6	4	5	3	9	8	1
5	6	1	2	3	9	8	7	4
7	8	9	6	4	1	2	5	3
4	3	2	5	7	8	6	1	9

SOLUTION 102

8	2	3	1	4	9	5	6	7
7	6	4	3	5	8	9	2	1
5	1	9	7	6	2	3	4	8
1	7	6	4	3	5	2	8	9
2	3	8	6	9	1	7	5	4
4	9	5	8	2	7	1	3	6
6	4	1	2	7	3	8	9	5
3	5	7	9	8	6	4	1	2
9	8	2	5	1	4	6	7	3

SOLUTION 103

1	4	9	6	2	3	7	5	8
5	8	2	7	1	4	3	6	9
3	6	7	5	9	8	4	1	2
2	7	6	1	3	5	8	9	4
8	3	5	2	4	9	6	7	1
9	1	4	8	6	7	2	3	5
4	9	1	3	7	2	5	8	6
7	2	8	9	5	6	1	4	3
6	5	3	4	8	1	9	2	7

SOLUTION 104

3	5	6	2	7	1	9	4	8
4	2	1	5	8	9	7	6	3
7	9	8	6	3	4	2	5	1
9	3	4	1	2	5	6	8	7
2	8	5	4	6	7	3	1	9
6	1	7	3	9	8	4	2	5
8	4	9	7	1	2	5	3	6
5	7	3	8	4	6	1	9	2
1	6	2	9	5	3	8	7	4

SOLUTION 105

6	2	3	4	9	7	5	1	8
8	9	7	5	1	6	3	4	2
4	5	1	2	8	3	6	9	7
1	6	4	9	2	5	7	8	3
2	3	5	1	7	8	4	6	9
7	8	9	6	3	4	1	2	5
9	7	6	3	4	2	8	5	1
3	4	2	8	5	1	9	7	6
5	1	8	7	6	9	2	3	4

SOLUTION 106

3	5	9	7	8	2	1	4	6
1	6	4	3	5	9	2	8	7
8	7	2	4	1	6	5	9	3
4	3	5	9	2	7	6	1	8
2	1	8	5	6	3	4	7	9
6	9	7	1	4	8	3	2	5
9	8	6	2	3	4	7	5	1
5	2	3	8	7	1	9	6	4
7	4	1	6	9	5	8	3	2

SOLUTION 107

1	2	5	9	6	4	7	8	3
4	3	6	5	7	8	1	2	9
7	8	9	1	3	2	5	4	6
9	7	2	4	1	3	8	6	5
8	4	1	6	2	5	3	9	7
5	6	3	7	8	9	2	1	4
2	5	8	3	4	6	9	7	1
3	1	4	8	9	7	6	5	2
6	9	7	2	5	1	4	3	8

SOLUTION 108

8	9	2	7	4	6	5	1	3
5	7	4	8	3	1	6	9	2
3	6	1	2	5	9	8	4	7
1	4	9	5	2	7	3	8	6
7	3	6	1	8	4	2	5	9
2	5	8	6	9	3	4	7	1
9	1	3	4	6	8	7	2	5
6	8	5	9	7	2	1	3	4
4	2	7	3	1	5	9	6	8

SOLUTION 109

4	2	9	7	3	5	8	6	1
5	6	8	2	9	1	3	7	4
7	3	1	8	6	4	2	9	5
8	1	5	4	7	2	6	3	9
3	4	7	9	5	6	1	8	2
2	9	6	3	1	8	5	4	7
1	7	3	5	8	9	4	2	6
6	8	2	1	4	7	9	5	3
9	5	4	6	2	3	7	1	8

SOLUTION 110

3	8	9	4	7	2	5	6	1
2	5	6	3	1	8	9	7	4
1	7	4	6	5	9	2	8	3
7	4	2	5	9	6	1	3	8
9	3	8	7	2	1	6	4	5
5	6	1	8	4	3	7	9	2
4	1	5	9	3	7	8	2	6
6	9	3	2	8	5	4	1	7
8	2	7	1	6	4	3	5	9

SOLUTION 111

9	5	1	3	8	4	7	6	2
6	7	2	5	1	9	8	3	4
3	4	8	7	2	6	5	1	9
8	1	6	2	7	3	4	9	5
4	2	7	9	5	1	3	8	6
5	9	3	6	4	8	1	2	7
1	6	5	4	3	2	9	7	8
7	3	9	8	6	5	2	4	1
2	8	4	1	9	7	6	5	3

SOLUTION 112

8	6	5	2	1	9	4	3	7
1	3	2	6	4	7	5	9	8
7	9	4	5	8	3	1	6	2
4	2	6	8	3	1	9	7	5
5	8	3	9	7	2	6	1	4
9	7	1	4	6	5	8	2	3
3	1	9	7	5	4	2	8	6
6	5	7	1	2	8	3	4	9
2	4	8	3	9	6	7	5	1

SOLUTION 113

2	8	1	5	9	4	7	3	6
9	7	5	1	6	3	4	8	2
3	4	6	8	2	7	9	1	5
6	3	2	4	1	9	5	7	8
7	5	8	2	3	6	1	4	9
1	9	4	7	8	5	6	2	3
4	2	3	6	5	1	8	9	7
5	1	9	3	7	8	2	6	4
8	6	7	9	4	2	3	5	1

SOLUTION 114

7	5	8	4	3	2	9	6	1
3	9	6	7	5	1	2	8	4
2	1	4	8	9	6	3	5	7
5	4	7	2	6	3	1	9	8
9	8	2	5	1	4	7	3	6
1	6	3	9	7	8	4	2	5
6	3	5	1	2	7	8	4	9
4	7	9	3	8	5	6	1	2
8	2	1	6	4	9	5	7	3

SOLUTION 115

2	6	8	4	3	5	9	7	1
3	7	5	9	8	1	2	4	6
1	9	4	7	6	2	8	3	5
8	1	9	5	7	3	6	2	4
5	3	6	2	4	8	7	1	9
4	2	7	6	1	9	5	8	3
9	5	1	8	2	4	3	6	7
7	4	2	3	9	6	1	5	8
6	8	3	1	5	7	4	9	2

SOLUTION 116

5	8	1	3	9	2	6	7	4
6	9	2	7	5	4	3	1	8
4	3	7	1	6	8	5	2	9
1	5	8	6	3	7	4	9	2
2	4	9	8	1	5	7	3	6
3	7	6	2	4	9	1	8	5
9	6	3	5	2	1	8	4	7
8	1	4	9	7	6	2	5	3
7	2	5	4	8	3	9	6	1

SOLUTION 117

2	3	4	8	6	5	7	9	1
7	8	5	4	9	1	2	6	3
6	1	9	3	2	7	8	4	5
8	5	6	9	1	2	3	7	4
4	9	3	5	7	8	6	1	2
1	7	2	6	4	3	5	8	9
5	2	1	7	8	9	4	3	6
9	6	7	2	3	4	1	5	8
3	4	8	1	5	6	9	2	7

SOLUTION 118

3	4	8	7	5	6	1	2	9
7	6	1	9	4	2	8	5	3
5	9	2	3	1	8	4	7	6
6	2	4	1	8	9	7	3	5
8	7	5	6	2	3	9	4	1
1	3	9	5	7	4	6	8	2
9	1	7	4	3	5	2	6	8
2	5	6	8	9	7	3	1	4
4	8	3	2	6	1	5	9	7

SOLUTION 119

9	3	1	2	8	5	6	7	4
7	4	6	3	1	9	2	8	5
8	2	5	7	4	6	3	1	9
6	1	4	5	3	7	9	2	8
2	5	7	9	6	8	4	3	1
3	9	8	1	2	4	5	6	7
1	6	9	4	7	3	8	5	2
4	7	3	8	5	2	1	9	6
5	8	2	6	9	1	7	4	3

SOLUTION 120

1	2	8	7	6	3	5	4	9
6	4	7	1	5	9	3	8	2
5	3	9	4	8	2	7	1	6
7	1	6	3	4	5	9	2	8
2	5	3	8	9	6	4	7	1
9	8	4	2	1	7	6	3	5
3	7	5	6	2	1	8	9	4
8	6	2	9	7	4	1	5	3
4	9	1	5	3	8	2	6	7

SOLUTION 121

9	2	6	5	7	1	4	3	8
7	8	5	2	4	3	1	6	9
1	4	3	6	8	9	2	5	7
8	3	4	9	6	7	5	1	2
2	1	9	3	5	4	7	8	6
6	5	7	1	2	8	3	9	4
5	7	1	4	9	6	8	2	3
4	6	2	8	3	5	9	7	1
3	9	8	7	1	2	6	4	5

SOLUTION 122

1	8	3	4	2	5	6	7	9
4	6	9	7	1	8	2	3	5
7	2	5	3	6	9	4	8	1
6	4	8	9	5	1	7	2	3
2	9	7	8	3	4	1	5	6
5	3	1	2	7	6	9	4	8
8	5	2	6	9	7	3	1	4
3	1	6	5	4	2	8	9	7
9	7	4	1	8	3	5	6	2

SOLUTION 123

5	6	1	4	7	9	3	8	2
8	9	2	3	5	6	1	4	7
4	7	3	8	1	2	5	9	6
7	2	6	5	3	8	4	1	9
9	1	5	6	4	7	2	3	8
3	8	4	2	9	1	6	7	5
1	5	7	9	2	3	8	6	4
6	4	9	1	8	5	7	2	3
2	3	8	7	6	4	9	5	1

SOLUTION 124

7	2	9	3	8	1	5	6	4
6	1	3	2	5	4	8	9	7
8	5	4	7	6	9	2	3	1
9	7	2	8	4	3	1	5	6
5	4	1	9	7	6	3	2	8
3	6	8	1	2	5	4	7	9
1	8	6	5	9	2	7	4	3
2	9	7	4	3	8	6	1	5
4	3	5	6	1	7	9	8	2

SOLUTION 125

4	6	8	9	7	1	2	5	3
7	1	2	4	3	5	8	9	6
3	5	9	6	8	2	7	1	4
5	8	4	2	6	9	3	7	1
6	2	3	5	1	7	4	8	9
1	9	7	8	4	3	6	2	5
2	3	6	1	9	8	5	4	7
8	7	1	3	5	4	9	6	2
9	4	5	7	2	6	1	3	8

SOLUTION 126

2	6	4	5	8	1	9	7	3
9	1	3	4	2	7	8	6	5
7	8	5	6	3	9	4	1	2
6	5	7	1	9	2	3	8	4
4	9	8	3	5	6	1	2	7
1	3	2	8	7	4	6	5	9
8	4	9	2	6	5	7	3	1
3	2	1	7	4	8	5	9	6
5	7	6	9	1	3	2	4	8

SOLUTION 127

3	9	6	4	7	8	1	5	2
5	7	2	1	3	9	4	6	8
8	1	4	6	5	2	9	3	7
1	3	5	8	6	4	7	2	9
4	6	8	2	9	7	3	1	5
9	2	7	5	1	3	8	4	6
2	5	3	7	8	1	6	9	4
6	8	9	3	4	5	2	7	1
7	4	1	9	2	6	5	8	3

SOLUTION 128

2	6	4	1	5	9	8	7	3
8	9	7	4	2	3	5	1	6
5	3	1	8	7	6	2	9	4
9	1	5	3	6	2	7	4	8
7	2	8	5	1	4	3	6	9
3	4	6	7	9	8	1	5	2
6	7	3	2	4	1	9	8	5
1	8	9	6	3	5	4	2	7
4	5	2	9	8	7	6	3	1

SOLUTION 129

8	5	1	6	3	9	7	2	4
2	3	6	5	7	4	1	8	9
9	4	7	1	8	2	6	3	5
7	6	3	4	2	8	5	9	1
5	9	4	3	1	7	2	6	8
1	8	2	9	6	5	3	4	7
3	1	8	7	9	6	4	5	2
6	2	5	8	4	1	9	7	3
4	7	9	2	5	3	8	1	6

SOLUTION 130

2	1	3	8	5	4	7	6	9
4	5	6	2	7	9	1	3	8
7	9	8	3	6	1	4	2	5
5	7	4	9	1	2	3	8	6
9	3	2	6	4	8	5	1	7
8	6	1	7	3	5	2	9	4
6	8	5	4	2	3	9	7	1
1	2	9	5	8	7	6	4	3
3	4	7	1	9	6	8	5	2

SOLUTION 131

6	2	5	7	9	3	1	8	4
9	3	8	4	2	1	5	7	6
4	1	7	8	5	6	3	2	9
1	4	2	5	6	9	7	3	8
8	5	3	1	7	4	6	9	2
7	9	6	2	3	8	4	1	5
2	8	4	6	1	7	9	5	3
3	6	1	9	8	5	2	4	7
5	7	9	3	4	2	8	6	1

SOLUTION 132

1	6	5	3	7	2	4	8	9
3	7	8	9	4	1	2	6	5
4	2	9	5	8	6	1	3	7
7	9	3	8	6	4	5	1	2
5	1	2	7	3	9	8	4	6
8	4	6	2	1	5	9	7	3
6	5	4	1	2	7	3	9	8
2	8	1	6	9	3	7	5	4
9	3	7	4	5	8	6	2	1

SOLUTION 133

1	8	5	3	6	4	2	9	7
9	3	7	5	2	1	4	8	6
2	4	6	9	7	8	1	3	5
5	6	1	4	8	9	7	2	3
3	2	8	1	5	7	9	6	4
7	9	4	6	3	2	8	5	1
8	1	3	7	9	6	5	4	2
6	7	2	8	4	5	3	1	9
4	5	9	2	1	3	6	7	8

SOLUTION 134

6	8	5	3	2	1	7	4	9
3	9	1	7	8	4	2	5	6
2	4	7	9	6	5	1	8	3
4	3	2	6	7	8	9	1	5
9	1	6	5	4	2	3	7	8
7	5	8	1	3	9	6	2	4
1	2	3	4	5	6	8	9	7
5	7	9	8	1	3	4	6	2
8	6	4	2	9	7	5	3	1

SOLUTION 135

7	2	1	4	5	9	3	6	8
4	3	8	1	7	6	2	9	5
9	5	6	2	8	3	4	7	1
2	6	3	9	1	5	7	8	4
5	8	4	3	6	7	9	1	2
1	9	7	8	4	2	6	5	3
6	4	5	7	3	1	8	2	9
8	7	2	5	9	4	1	3	6
3	1	9	6	2	8	5	4	7

SOLUTION 136

9	8	3	5	1	4	6	2	7
4	6	2	8	3	7	1	5	9
1	5	7	9	6	2	3	8	4
6	2	9	4	5	8	7	1	3
8	7	5	3	9	1	2	4	6
3	4	1	7	2	6	8	9	5
7	1	4	6	8	5	9	3	2
5	3	8	2	7	9	4	6	1
2	9	6	1	4	3	5	7	8

SOLUTION 137

2	3	6	5	8	7	1	4	9
9	4	7	3	2	1	8	5	6
1	8	5	4	6	9	3	2	7
3	9	1	8	7	5	2	6	4
7	5	8	6	4	2	9	1	3
4	6	2	1	9	3	5	7	8
6	2	3	9	1	4	7	8	5
8	7	9	2	5	6	4	3	1
5	1	4	7	3	8	6	9	2

SOLUTION 138

8	9	1	3	2	7	6	4	5
4	5	6	1	8	9	7	2	3
2	3	7	4	5	6	8	1	9
6	4	3	7	1	2	9	5	8
5	7	2	8	9	3	4	6	1
1	8	9	5	6	4	2	3	7
9	1	5	2	4	8	3	7	6
7	2	8	6	3	5	1	9	4
3	6	4	9	7	1	5	8	2

SOLUTION 139

5	3	6	7	2	1	9	4	8
1	9	2	8	3	4	7	6	5
7	8	4	5	6	9	3	2	1
8	7	3	2	1	6	5	9	4
6	4	1	9	5	8	2	3	7
2	5	9	3	4	7	1	8	6
4	2	5	6	7	3	8	1	9
9	1	7	4	8	2	6	5	3
3	6	8	1	9	5	4	7	2

SOLUTION 140

2	6	4	3	1	9	8	7	5
9	5	3	2	8	7	6	4	1
1	8	7	4	6	5	9	3	2
3	4	1	6	9	8	5	2	7
6	7	2	1	5	4	3	9	8
5	9	8	7	2	3	1	6	4
8	1	6	9	4	2	7	5	3
7	2	9	5	3	1	4	8	6
4	3	5	8	7	6	2	1	9

SOLUTION 141

2	4	5	8	9	3	1	7	6
1	6	9	7	4	2	3	5	8
7	3	8	6	1	5	9	4	2
4	5	2	3	6	9	8	1	7
8	9	1	5	2	7	6	3	4
3	7	6	1	8	4	5	2	9
5	2	4	9	3	6	7	8	1
9	1	7	2	5	8	4	6	3
6	8	3	4	7	1	2	9	5

SOLUTION 142

7	1	9	4	8	6	2	3	5
2	6	4	9	5	3	1	8	7
5	3	8	1	2	7	9	6	4
6	5	3	2	7	4	8	1	9
4	9	7	6	1	8	5	2	3
1	8	2	3	9	5	4	7	6
8	7	1	5	3	9	6	4	2
9	2	6	7	4	1	3	5	8
3	4	5	8	6	2	7	9	1

SOLUTION 143

2	3	4	6	1	5	8	7	9
5	6	7	3	9	8	1	2	4
1	9	8	4	2	7	3	5	6
6	7	5	2	4	3	9	1	8
3	8	9	7	5	1	4	6	2
4	2	1	9	8	6	5	3	7
8	1	2	5	7	4	6	9	3
7	5	6	8	3	9	2	4	1
9	4	3	1	6	2	7	8	5

SOLUTION 144

5	4	2	9	6	8	1	3	7
1	9	3	7	4	5	8	2	6
6	7	8	3	1	2	5	4	9
3	6	4	1	8	9	2	7	5
9	1	7	5	2	4	6	8	3
2	8	5	6	3	7	9	1	4
8	2	6	4	9	3	7	5	1
4	5	1	8	7	6	3	9	2
7	3	9	2	5	1	4	6	8

SOLUTION 145

8	7	1	4	5	6	3	9	2
5	3	6	1	2	9	4	8	7
4	2	9	7	3	8	6	5	1
7	9	2	6	8	5	1	4	3
6	4	5	2	1	3	9	7	8
3	1	8	9	7	4	5	2	6
1	8	3	5	9	7	2	6	4
2	5	4	8	6	1	7	3	9
9	6	7	3	4	2	8	1	5

SOLUTION 146

6	4	7	9	3	1	5	2	8
9	2	1	6	8	5	4	7	3
8	3	5	2	7	4	6	1	9
5	6	8	1	2	7	9	3	4
7	1	4	3	9	8	2	5	6
3	9	2	5	4	6	1	8	7
2	8	3	4	1	9	7	6	5
1	5	9	7	6	3	8	4	2
4	7	6	8	5	2	3	9	1

SOLUTION 147

3	2	7	6	9	1	4	8	5
9	5	1	4	3	8	7	2	6
8	6	4	5	7	2	3	9	1
4	3	8	7	5	6	2	1	9
5	1	6	2	4	9	8	7	3
7	9	2	1	8	3	5	6	4
2	4	3	9	6	7	1	5	8
1	8	9	3	2	5	6	4	7
6	7	5	8	1	4	9	3	2

SOLUTION 148

2	4	7	6	5	8	9	3	1
3	9	5	7	1	4	8	2	6
8	1	6	2	3	9	7	4	5
5	3	8	1	6	7	4	9	2
4	6	1	3	9	2	5	7	8
9	7	2	4	8	5	1	6	3
6	8	4	5	7	3	2	1	9
7	5	3	9	2	1	6	8	4
1	2	9	8	4	6	3	5	7

SOLUTION 149

2	9	3	4	8	1	7	5	6
1	4	5	9	7	6	3	8	2
7	8	6	2	5	3	4	9	1
9	5	8	6	1	7	2	3	4
3	7	1	8	4	2	5	6	9
4	6	2	3	9	5	8	1	7
6	1	7	5	2	8	9	4	3
5	2	4	1	3	9	6	7	8
8	3	9	7	6	4	1	2	5

SOLUTION 150

7	2	5	3	9	4	6	1	8
4	1	3	7	8	6	5	2	9
8	9	6	1	5	2	7	3	4
2	4	7	6	3	5	8	9	1
9	5	1	4	2	8	3	6	7
6	3	8	9	1	7	4	5	2
3	7	4	2	6	9	1	8	5
5	6	9	8	4	1	2	7	3
1	8	2	5	7	3	9	4	6

SOLUTION 151

1	2	4	3	9	5	7	6	8
6	3	5	4	7	8	1	9	2
9	7	8	6	2	1	5	3	4
4	5	2	1	6	7	9	8	3
3	6	9	8	5	4	2	7	1
8	1	7	9	3	2	4	5	6
7	4	1	5	8	3	6	2	9
2	9	3	7	4	6	8	1	5
5	8	6	2	1	9	3	4	7

SOLUTION 152

9	5	4	7	3	1	2	8	6
7	3	8	6	9	2	4	1	5
2	6	1	8	5	4	7	3	9
5	4	2	9	1	7	3	6	8
1	7	6	5	8	3	9	2	4
8	9	3	4	2	6	5	7	1
6	1	5	2	7	9	8	4	3
3	2	9	1	4	8	6	5	7
4	8	7	3	6	5	1	9	2

SOLUTION 153

2	3	1	4	6	5	8	7	9
9	6	7	8	3	2	5	4	1
8	4	5	9	7	1	2	3	6
5	8	6	7	2	3	1	9	4
7	1	4	6	5	9	3	2	8
3	9	2	1	4	8	7	6	5
4	2	9	5	8	7	6	1	3
6	7	8	3	1	4	9	5	2
1	5	3	2	9	6	4	8	7

SOLUTION 154

4	2	3	6	7	8	1	5	9
7	6	1	9	3	5	2	8	4
9	8	5	1	2	4	3	6	7
1	9	4	7	6	2	5	3	8
6	3	8	4	5	1	9	7	2
5	7	2	3	8	9	4	1	6
3	4	7	5	9	6	8	2	1
8	5	9	2	1	7	6	4	3
2	1	6	8	4	3	7	9	5

SOLUTION 155

9	6	5	7	1	8	2	4	3
1	3	4	5	2	9	8	6	7
2	8	7	6	4	3	5	1	9
6	4	8	3	5	1	7	9	2
5	1	3	2	9	7	4	8	6
7	2	9	8	6	4	1	3	5
4	7	6	1	3	5	9	2	8
8	9	2	4	7	6	3	5	1
3	5	1	9	8	2	6	7	4

SOLUTION 156

9	8	5	7	2	1	4	3	6
4	7	3	5	8	6	9	1	2
6	1	2	4	3	9	5	7	8
1	9	4	3	7	2	6	8	5
2	5	8	1	6	4	3	9	7
7	3	6	8	9	5	1	2	4
5	6	9	2	1	7	8	4	3
8	4	7	9	5	3	2	6	1
3	2	1	6	4	8	7	5	9

SOLUTION 157

4	6	3	5	9	8	2	7	1
5	2	9	7	4	1	6	3	8
8	1	7	3	2	6	9	4	5
6	9	2	8	5	7	3	1	4
3	8	1	2	6	4	7	5	9
7	5	4	9	1	3	8	2	6
2	4	6	1	3	9	5	8	7
1	7	5	6	8	2	4	9	3
9	3	8	4	7	5	1	6	2

SOLUTION 158

2	7	4	8	1	6	3	5	9
6	8	9	4	3	5	1	2	7
3	1	5	2	9	7	6	8	4
5	3	1	7	2	8	4	9	6
4	2	7	6	5	9	8	3	1
9	6	8	3	4	1	5	7	2
1	4	3	9	8	2	7	6	5
7	5	2	1	6	3	9	4	8
8	9	6	5	7	4	2	1	3

SOLUTION 159

4	5	3	8	6	2	7	1	9
6	8	1	3	7	9	4	2	5
9	7	2	5	1	4	3	6	8
3	9	4	7	2	6	8	5	1
2	6	5	1	8	3	9	7	4
8	1	7	4	9	5	6	3	2
1	2	6	9	4	7	5	8	3
7	3	9	2	5	8	1	4	6
5	4	8	6	3	1	2	9	7

SOLUTION 160

8	1	2	6	5	4	3	7	9
7	3	5	2	9	1	4	6	8
6	9	4	3	7	8	1	2	5
1	6	8	5	2	9	7	3	4
3	2	9	1	4	7	5	8	6
5	4	7	8	3	6	2	9	1
4	7	1	9	6	3	8	5	2
2	8	6	7	1	5	9	4	3
9	5	3	4	8	2	6	1	7

SOLUTION 161

4	7	3	9	2	6	1	5	8
8	5	6	3	1	4	2	9	7
9	2	1	5	8	7	6	4	3
5	9	7	1	4	2	3	8	6
2	3	4	6	5	8	7	1	9
6	1	8	7	3	9	4	2	5
1	8	5	4	6	3	9	7	2
3	4	9	2	7	5	8	6	1
7	6	2	8	9	1	5	3	4

SOLUTION 162

2	5	9	3	1	8	6	7	4
6	1	7	9	5	4	3	2	8
4	3	8	7	2	6	5	9	1
9	4	1	8	6	7	2	3	5
5	6	3	2	4	9	8	1	7
7	8	2	1	3	5	4	6	9
3	9	5	6	8	1	7	4	2
8	7	6	4	9	2	1	5	3
1	2	4	5	7	3	9	8	6

SOLUTION 163

9	2	5	3	6	7	1	4	8
3	6	7	1	4	8	2	9	5
4	8	1	5	2	9	7	3	6
5	3	2	9	1	6	4	8	7
1	4	8	2	7	3	5	6	9
7	9	6	8	5	4	3	2	1
6	5	3	4	8	1	9	7	2
8	1	4	7	9	2	6	5	3
2	7	9	6	3	5	8	1	4

SOLUTION 164

1	4	7	6	8	9	5	2	3
3	5	6	1	4	2	7	8	9
8	9	2	3	7	5	6	4	1
5	7	1	2	9	3	4	6	8
9	6	8	4	5	7	1	3	2
4	2	3	8	1	6	9	7	5
7	3	5	9	2	4	8	1	6
6	1	4	5	3	8	2	9	7
2	8	9	7	6	1	3	5	4

SOLUTION 165

1	2	4	9	6	5	3	7	8
8	3	7	2	4	1	5	6	9
6	9	5	7	3	8	4	1	2
4	7	6	1	8	2	9	3	5
2	8	9	5	7	3	6	4	1
3	5	1	4	9	6	8	2	7
7	1	3	6	5	9	2	8	4
9	6	2	8	1	4	7	5	3
5	4	8	3	2	7	1	9	6

SOLUTION 166

3	7	9	2	1	8	5	4	6
5	2	4	6	7	9	1	3	8
6	1	8	4	3	5	9	7	2
2	3	5	1	9	6	4	8	7
1	4	7	5	8	3	2	6	9
9	8	6	7	2	4	3	5	1
7	5	3	9	6	2	8	1	4
8	6	2	3	4	1	7	9	5
4	9	1	8	5	7	6	2	3

SOLUTION 167

7	5	9	3	2	8	1	6	4
8	2	3	1	4	6	7	9	5
6	1	4	5	9	7	8	2	3
4	6	2	8	5	9	3	7	1
9	3	8	7	6	1	5	4	2
5	7	1	4	3	2	9	8	6
1	4	7	2	8	5	6	3	9
2	9	5	6	7	3	4	1	8
3	8	6	9	1	4	2	5	7

SOLUTION 168

3	2	6	8	1	9	5	4	7
9	5	4	2	7	3	8	6	1
8	1	7	5	4	6	2	3	9
4	3	2	9	5	1	6	7	8
7	9	8	3	6	2	4	1	5
1	6	5	7	8	4	9	2	3
5	7	3	4	2	8	1	9	6
2	8	1	6	9	7	3	5	4
6	4	9	1	3	5	7	8	2

SOLUTION 169

2	1	5	3	9	7	8	4	6
9	6	4	8	2	1	5	3	7
3	8	7	6	5	4	1	2	9
6	5	3	2	1	9	4	7	8
7	2	8	4	6	3	9	1	5
1	4	9	7	8	5	3	6	2
4	3	2	9	7	8	6	5	1
5	9	6	1	4	2	7	8	3
8	7	1	5	3	6	2	9	4

SOLUTION 170

9	8	7	3	1	5	4	6	2
5	3	4	2	8	6	9	1	7
1	6	2	7	9	4	3	8	5
8	7	3	4	6	2	5	9	1
4	2	1	9	5	3	6	7	8
6	9	5	1	7	8	2	4	3
2	4	8	6	3	7	1	5	9
7	1	6	5	2	9	8	3	4
3	5	9	8	4	1	7	2	6

SOLUTION 171

2	5	9	3	4	6	8	7	1
6	3	4	7	1	8	9	2	5
7	1	8	5	2	9	3	6	4
3	7	6	8	5	4	1	9	2
5	4	2	1	9	3	6	8	7
8	9	1	6	7	2	5	4	3
1	2	5	9	6	7	4	3	8
4	6	3	2	8	5	7	1	9
9	8	7	4	3	1	2	5	6

SOLUTION 172

2	8	1	9	7	6	4	3	5
6	5	9	8	4	3	2	1	7
7	4	3	1	2	5	6	9	8
9	7	4	3	1	2	5	8	6
1	3	2	6	5	8	9	7	4
8	6	5	4	9	7	3	2	1
3	9	7	5	8	4	1	6	2
4	2	6	7	3	1	8	5	9
5	1	8	2	6	9	7	4	3

SOLUTION 173

4	8	5	7	9	3	1	6	2
6	2	3	4	5	1	7	9	8
1	7	9	6	2	8	4	3	5
8	1	4	3	7	2	6	5	9
3	9	7	1	6	5	2	8	4
5	6	2	9	8	4	3	1	7
7	5	6	2	3	9	8	4	1
9	3	1	8	4	7	5	2	6
2	4	8	5	1	6	9	7	3

SOLUTION 174

3	8	4	9	7	2	1	5	6
2	9	7	5	1	6	4	3	8
5	6	1	8	3	4	2	7	9
6	4	5	1	8	7	9	2	3
9	1	8	3	2	5	7	6	4
7	3	2	4	6	9	8	1	5
4	2	3	6	9	1	5	8	7
1	5	6	7	4	8	3	9	2
8	7	9	2	5	3	6	4	1

SOLUTION 175

3	2	6	8	9	5	7	4	1
7	1	9	4	2	3	6	8	5
5	8	4	6	7	1	9	2	3
2	9	1	7	5	6	4	3	8
6	4	3	1	8	9	2	5	7
8	5	7	2	3	4	1	6	9
1	6	5	3	4	7	8	9	2
9	7	2	5	6	8	3	1	4
4	3	8	9	1	2	5	7	6

SOLUTION 176

1	2	6	3	4	9	8	7	5
4	9	3	7	8	5	6	1	2
8	5	7	6	2	1	4	3	9
2	1	8	9	7	6	3	5	4
7	3	5	8	1	4	9	2	6
6	4	9	5	3	2	7	8	1
3	6	2	4	5	7	1	9	8
5	8	4	1	9	3	2	6	7
9	7	1	2	6	8	5	4	3

SOLUTION 177

9	4	5	8	3	2	1	7	6
2	8	1	9	6	7	5	4	3
7	3	6	1	4	5	2	9	8
5	9	3	2	8	1	4	6	7
1	2	4	3	7	6	9	8	5
6	7	8	4	5	9	3	1	2
8	5	2	6	1	4	7	3	9
3	1	7	5	9	8	6	2	4
4	6	9	7	2	3	8	5	1

SOLUTION 178

3	5	9	4	7	8	6	1	2
4	2	8	9	6	1	7	5	3
1	6	7	3	5	2	9	8	4
5	4	3	2	1	7	8	6	9
2	9	1	6	8	4	5	3	7
8	7	6	5	9	3	4	2	1
7	1	4	8	2	5	3	9	6
6	8	2	7	3	9	1	4	5
9	3	5	1	4	6	2	7	8

SOLUTION 179

6	4	3	9	2	5	1	7	8
9	5	7	1	4	8	6	3	2
2	1	8	3	6	7	9	5	4
4	3	6	8	5	1	2	9	7
1	7	9	6	3	2	4	8	5
8	2	5	7	9	4	3	6	1
7	6	1	4	8	3	5	2	9
5	9	4	2	7	6	8	1	3
3	8	2	5	1	9	7	4	6

SOLUTION 180

9	3	1	7	8	6	5	2	4
7	4	6	2	3	5	1	9	8
2	8	5	9	4	1	6	7	3
1	9	7	6	2	8	4	3	5
3	5	4	1	9	7	2	8	6
6	2	8	4	5	3	7	1	9
4	1	3	5	7	9	8	6	2
5	6	9	8	1	2	3	4	7
8	7	2	3	6	4	9	5	1

SOLUTION 181

5	2	1	6	9	3	4	7	8
7	8	6	4	5	2	3	1	9
9	4	3	1	7	8	2	5	6
8	5	7	3	4	9	1	6	2
1	6	9	8	2	7	5	3	4
2	3	4	5	1	6	9	8	7
6	9	2	7	3	1	8	4	5
3	7	5	9	8	4	6	2	1
4	1	8	2	6	5	7	9	3

SOLUTION 182

4	1	9	3	2	6	7	8	5
2	8	6	1	5	7	3	4	9
7	3	5	9	4	8	1	2	6
5	2	3	4	9	1	8	6	7
9	4	7	6	8	2	5	1	3
1	6	8	7	3	5	2	9	4
6	5	4	8	1	3	9	7	2
3	9	1	2	7	4	6	5	8
8	7	2	5	6	9	4	3	1

SOLUTION 183

8	1	3	5	7	4	9	2	6
6	5	7	1	9	2	3	4	8
9	4	2	3	6	8	7	1	5
2	6	9	4	3	1	5	8	7
5	3	4	2	8	7	6	9	1
1	7	8	6	5	9	2	3	4
4	2	5	7	1	3	8	6	9
3	9	6	8	4	5	1	7	2
7	8	1	9	2	6	4	5	3

SOLUTION 184

1	5	4	9	3	6	2	7	8
2	6	7	4	5	8	3	9	1
3	8	9	7	1	2	4	6	5
6	4	3	5	8	1	7	2	9
5	7	2	6	9	4	8	1	3
9	1	8	2	7	3	5	4	6
4	2	1	3	6	5	9	8	7
8	9	5	1	2	7	6	3	4
7	3	6	8	4	9	1	5	2

SOLUTION 185

4	8	3	2	1	5	7	6	9
6	2	1	4	7	9	8	5	3
5	9	7	6	8	3	4	2	1
7	4	5	1	2	8	3	9	6
9	6	8	3	5	4	1	7	2
3	1	2	7	9	6	5	8	4
8	7	4	9	6	1	2	3	5
2	3	6	5	4	7	9	1	8
1	5	9	8	3	2	6	4	7

SOLUTION 186

4	6	2	7	9	8	3	5	1
3	5	8	6	2	1	9	4	7
1	9	7	3	5	4	6	8	2
2	3	4	1	6	7	8	9	5
8	1	5	2	3	9	4	7	6
9	7	6	8	4	5	2	1	3
6	4	1	5	8	2	7	3	9
7	2	9	4	1	3	5	6	8
5	8	3	9	7	6	1	2	4

SOLUTION 187

3	5	9	1	4	8	2	6	7
1	4	2	3	6	7	5	9	8
7	8	6	2	5	9	3	4	1
6	1	5	8	2	4	7	3	9
2	3	4	9	7	5	8	1	6
8	9	7	6	3	1	4	5	2
5	7	1	4	9	2	6	8	3
4	6	8	7	1	3	9	2	5
9	2	3	5	8	6	1	7	4

SOLUTION 188

6	3	4	5	9	7	2	8	1
8	1	9	6	3	2	5	7	4
7	2	5	8	4	1	9	3	6
2	6	3	4	5	8	1	9	7
4	9	7	1	6	3	8	5	2
5	8	1	7	2	9	6	4	3
3	5	8	2	1	4	7	6	9
1	4	6	9	7	5	3	2	8
9	7	2	3	8	6	4	1	5

SOLUTION 189

9	2	8	3	1	7	4	6	5
6	4	7	5	2	8	9	3	1
3	1	5	9	6	4	2	7	8
4	8	3	7	9	1	5	2	6
5	9	6	8	3	2	1	4	7
1	7	2	4	5	6	3	8	9
2	5	4	6	8	9	7	1	3
7	6	9	1	4	3	8	5	2
8	3	1	2	7	5	6	9	4

SOLUTION 190

3	2	7	8	6	1	4	9	5
8	6	1	4	5	9	3	2	7
9	5	4	3	7	2	6	8	1
2	9	5	6	1	4	8	7	3
1	3	6	5	8	7	9	4	2
7	4	8	9	2	3	1	5	6
6	8	2	1	9	5	7	3	4
4	7	9	2	3	6	5	1	8
5	1	3	7	4	8	2	6	9

SOLUTION 191

3	7	1	8	4	9	6	2	5
5	4	9	6	2	3	8	7	1
6	8	2	5	1	7	9	4	3
8	2	3	7	6	4	1	5	9
9	6	4	1	5	8	2	3	7
1	5	7	3	9	2	4	6	8
4	3	8	2	7	1	5	9	6
2	1	5	9	3	6	7	8	4
7	9	6	4	8	5	3	1	2

SOLUTION 192

8	3	6	7	5	2	1	4	9
1	4	9	8	6	3	2	7	5
2	5	7	4	9	1	8	6	3
3	1	4	5	7	6	9	2	8
6	2	5	3	8	9	4	1	7
7	9	8	1	2	4	3	5	6
4	8	1	6	3	5	7	9	2
9	6	3	2	4	7	5	8	1
5	7	2	9	1	8	6	3	4

SOLUTION 193

3	1	5	4	2	8	6	7	9
2	9	4	5	6	7	3	8	1
6	7	8	9	3	1	2	5	4
7	3	6	8	9	5	4	1	2
9	8	1	2	7	4	5	6	3
5	4	2	3	1	6	8	9	7
1	2	3	6	5	9	7	4	8
4	5	9	7	8	2	1	3	6
8	6	7	1	4	3	9	2	5

SOLUTION 194

2	9	6	4	7	8	5	3	1
1	5	4	2	3	9	8	7	6
3	8	7	5	6	1	9	4	2
7	6	3	9	2	5	1	8	4
4	2	5	1	8	3	7	6	9
8	1	9	6	4	7	2	5	3
5	7	2	3	9	4	6	1	8
9	4	1	8	5	6	3	2	7
6	3	8	7	1	2	4	9	5

SOLUTION 195

6	7	4	2	1	5	9	3	8
1	9	2	7	8	3	5	4	6
5	8	3	4	6	9	7	2	1
4	1	8	5	9	6	2	7	3
3	2	6	8	7	4	1	9	5
7	5	9	3	2	1	6	8	4
2	4	7	6	5	8	3	1	9
8	6	1	9	3	2	4	5	7
9	3	5	1	4	7	8	6	2

SOLUTION 196

6	2	3	7	5	4	1	9	8
7	8	4	2	9	1	6	3	5
5	9	1	6	3	8	7	2	4
4	3	9	8	2	7	5	6	1
1	5	6	9	4	3	2	8	7
8	7	2	1	6	5	9	4	3
3	6	8	5	1	9	4	7	2
9	4	5	3	7	2	8	1	6
2	1	7	4	8	6	3	5	9

SOLUTION 197

3	6	9	8	2	5	1	7	4
8	5	4	1	7	9	2	3	6
7	1	2	6	4	3	8	9	5
6	4	7	3	9	2	5	8	1
1	8	5	4	6	7	3	2	9
2	9	3	5	8	1	6	4	7
4	2	8	7	5	6	9	1	3
5	7	1	9	3	8	4	6	2
9	3	6	2	1	4	7	5	8

SOLUTION 198

4	6	8	2	3	5	9	1	7
9	7	2	8	1	4	6	3	5
5	1	3	6	7	9	4	8	2
6	9	7	1	2	3	8	5	4
8	3	4	9	5	6	2	7	1
1	2	5	7	4	8	3	6	9
7	5	6	3	9	2	1	4	8
3	4	9	5	8	1	7	2	6
2	8	1	4	6	7	5	9	3

SOLUTION 199

4	3	7	6	5	8	2	1	9
1	2	5	4	7	9	3	6	8
9	6	8	2	1	3	5	7	4
7	9	6	3	8	5	4	2	1
5	1	3	9	2	4	7	8	6
8	4	2	7	6	1	9	5	3
6	7	9	8	3	2	1	4	5
2	5	4	1	9	6	8	3	7
3	8	1	5	4	7	6	9	2

SOLUTION 200

2	1	9	8	4	7	5	6	3
7	8	6	5	3	1	4	2	9
5	4	3	9	6	2	1	7	8
3	2	7	4	5	6	8	9	1
4	9	8	1	2	3	7	5	6
1	6	5	7	8	9	3	4	2
9	7	4	2	1	8	6	3	5
6	5	1	3	9	4	2	8	7
8	3	2	6	7	5	9	1	4

SOLUTION 201

5	1	4	7	2	6	3	8	9
2	3	7	1	8	9	6	5	4
6	9	8	3	4	5	2	7	1
1	4	9	2	3	8	7	6	5
8	2	6	5	7	1	9	4	3
7	5	3	6	9	4	1	2	8
3	7	5	4	1	2	8	9	6
9	6	1	8	5	7	4	3	2
4	8	2	9	6	3	5	1	7

SOLUTION 202

5	2	1	3	6	9	4	7	8
7	8	6	1	5	4	2	9	3
9	4	3	2	8	7	1	6	5
2	5	4	7	3	1	9	8	6
1	3	9	6	2	8	7	5	4
8	6	7	4	9	5	3	1	2
4	7	2	8	1	6	5	3	9
6	1	5	9	4	3	8	2	7
3	9	8	5	7	2	6	4	1

SOLUTION 203

7	2	5	4	6	1	9	8	3
8	6	9	3	2	5	7	4	1
1	3	4	8	9	7	5	2	6
2	5	6	7	4	3	8	1	9
4	8	3	5	1	9	6	7	2
9	1	7	6	8	2	4	3	5
5	7	1	9	3	4	2	6	8
6	9	2	1	7	8	3	5	4
3	4	8	2	5	6	1	9	7

SOLUTION 204

3	6	7	4	5	1	9	8	2
8	1	4	9	2	3	5	7	6
2	5	9	7	8	6	4	3	1
4	8	5	6	7	9	1	2	3
6	9	1	3	4	2	8	5	7
7	3	2	8	1	5	6	9	4
1	4	8	2	9	7	3	6	5
5	7	6	1	3	8	2	4	9
9	2	3	5	6	4	7	1	8

SOLUTION 205

9	3	6	4	1	2	7	5	8
5	2	1	9	7	8	3	6	4
8	7	4	3	6	5	1	2	9
1	8	5	2	9	7	6	4	3
7	4	9	6	5	3	2	8	1
3	6	2	1	8	4	5	9	7
2	1	3	8	4	6	9	7	5
6	5	8	7	3	9	4	1	2
4	9	7	5	2	1	8	3	6

SOLUTION 206

4	5	9	3	2	6	7	1	8
8	3	7	1	5	4	6	2	9
2	1	6	7	8	9	4	5	3
7	8	3	6	1	5	2	9	4
5	4	1	9	3	2	8	6	7
6	9	2	4	7	8	5	3	1
1	6	5	8	4	3	9	7	2
3	2	4	5	9	7	1	8	6
9	7	8	2	6	1	3	4	5

SOLUTION 207

3	2	7	6	1	8	5	9	4
9	5	4	2	3	7	6	8	1
6	8	1	9	5	4	3	7	2
7	4	5	3	8	6	1	2	9
1	3	6	7	2	9	4	5	8
2	9	8	1	4	5	7	6	3
5	6	2	4	9	3	8	1	7
8	1	3	5	7	2	9	4	6
4	7	9	8	6	1	2	3	5

SOLUTION 208

1	6	4	7	2	8	3	5	9
3	9	5	1	4	6	8	7	2
8	7	2	3	9	5	4	6	1
6	3	9	2	7	4	5	1	8
7	5	1	6	8	9	2	3	4
2	4	8	5	3	1	7	9	6
9	8	7	4	6	3	1	2	5
4	1	3	9	5	2	6	8	7
5	2	6	8	1	7	9	4	3

SOLUTION 209

5	1	8	4	9	6	2	3	7
2	9	6	5	3	7	8	4	1
3	7	4	2	8	1	6	5	9
4	3	2	6	5	9	7	1	8
7	8	1	3	4	2	5	9	6
6	5	9	7	1	8	3	2	4
8	2	5	1	6	4	9	7	3
1	6	7	9	2	3	4	8	5
9	4	3	8	7	5	1	6	2

SOLUTION 210

3	9	1	2	6	5	7	8	4
2	6	4	9	8	7	5	3	1
7	8	5	1	3	4	2	6	9
4	2	3	6	5	9	1	7	8
5	7	9	4	1	8	3	2	6
8	1	6	7	2	3	9	4	5
6	5	7	8	9	2	4	1	3
1	3	2	5	4	6	8	9	7
9	4	8	3	7	1	6	5	2

SOLUTION 211

5	7	2	6	9	4	8	1	3
3	1	8	5	7	2	6	9	4
4	6	9	1	3	8	2	7	5
1	4	5	8	2	3	7	6	9
2	9	6	4	5	7	3	8	1
7	8	3	9	1	6	5	4	2
9	3	4	7	6	5	1	2	8
8	5	7	2	4	1	9	3	6
6	2	1	3	8	9	4	5	7

SOLUTION 212

1	6	9	4	3	2	8	5	7
2	8	3	7	9	5	4	6	1
7	4	5	6	1	8	2	3	9
5	3	6	2	7	1	9	4	8
8	2	4	5	6	9	7	1	3
9	7	1	8	4	3	5	2	6
3	5	7	9	2	6	1	8	4
4	1	2	3	8	7	6	9	5
6	9	8	1	5	4	3	7	2

SOLUTION 213

6	9	1	5	2	3	4	7	8
5	3	2	4	7	8	6	9	1
8	7	4	6	9	1	2	5	3
4	6	5	2	1	9	3	8	7
3	1	7	8	6	4	5	2	9
2	8	9	7	3	5	1	6	4
7	2	3	1	8	6	9	4	5
1	4	6	9	5	7	8	3	2
9	5	8	3	4	2	7	1	6

SOLUTION 214

6	7	3	8	2	4	5	1	9
8	2	9	5	1	3	4	6	7
5	1	4	7	6	9	8	2	3
1	6	7	3	8	5	2	9	4
3	4	5	9	7	2	1	8	6
9	8	2	1	4	6	3	7	5
7	5	6	2	3	1	9	4	8
4	3	1	6	9	8	7	5	2
2	9	8	4	5	7	6	3	1

SOLUTION 215

1	8	3	4	6	7	5	2	9
6	2	9	3	1	5	7	8	4
5	7	4	8	9	2	6	3	1
2	3	1	5	4	6	8	9	7
7	9	5	1	2	8	3	4	6
4	6	8	9	7	3	2	1	5
8	1	2	7	5	9	4	6	3
9	5	6	2	3	4	1	7	8
3	4	7	6	8	1	9	5	2

SOLUTION 216

1	7	2	5	4	3	8	9	6
8	9	5	6	7	1	3	2	4
3	6	4	2	8	9	5	7	1
6	8	3	1	2	5	9	4	7
5	4	9	8	6	7	1	3	2
2	1	7	9	3	4	6	5	8
4	3	1	7	9	6	2	8	5
7	5	8	3	1	2	4	6	9
9	2	6	4	5	8	7	1	3

SOLUTION 217

3	9	1	2	4	6	7	8	5
6	2	7	8	1	5	9	4	3
4	5	8	9	7	3	1	6	2
8	7	5	4	2	9	3	1	6
1	3	6	5	8	7	4	2	9
9	4	2	3	6	1	8	5	7
7	6	3	1	5	4	2	9	8
2	1	9	6	3	8	5	7	4
5	8	4	7	9	2	6	3	1

SOLUTION 218

4	1	9	7	8	5	6	2	3
8	2	6	4	3	1	9	7	5
7	3	5	2	6	9	4	1	8
2	4	1	9	5	7	3	8	6
3	9	8	6	2	4	7	5	1
6	5	7	8	1	3	2	4	9
5	6	3	1	7	2	8	9	4
1	7	4	3	9	8	5	6	2
9	8	2	5	4	6	1	3	7

SOLUTION 219

8	7	9	5	1	4	6	2	3
2	1	3	7	8	6	5	9	4
5	6	4	2	3	9	1	7	8
9	5	2	3	6	8	7	4	1
4	8	1	9	7	5	3	6	2
7	3	6	4	2	1	8	5	9
6	4	5	1	9	3	2	8	7
1	2	8	6	4	7	9	3	5
3	9	7	8	5	2	4	1	6

SOLUTION 220

9	5	8	4	1	3	2	6	7
3	1	4	6	7	2	5	9	8
2	6	7	5	9	8	3	1	4
6	4	1	3	8	9	7	2	5
8	2	9	7	5	6	4	3	1
7	3	5	2	4	1	6	8	9
1	7	6	8	3	5	9	4	2
4	9	3	1	2	7	8	5	6
5	8	2	9	6	4	1	7	3

SOLUTION 221

5	3	6	7	4	2	8	1	9
7	2	4	1	8	9	3	5	6
8	1	9	5	3	6	2	4	7
2	6	8	3	1	5	9	7	4
9	5	7	6	2	4	1	3	8
1	4	3	8	9	7	5	6	2
4	7	2	9	5	3	6	8	1
6	8	5	2	7	1	4	9	3
3	9	1	4	6	8	7	2	5

SOLUTION 222

2	3	5	4	6	9	1	8	7
4	8	9	3	1	7	2	6	5
7	1	6	2	5	8	4	9	3
3	7	8	5	2	4	9	1	6
6	9	1	8	7	3	5	2	4
5	2	4	6	9	1	7	3	8
1	4	2	7	3	6	8	5	9
9	6	7	1	8	5	3	4	2
8	5	3	9	4	2	6	7	1

SOLUTION 223

9	6	2	1	8	7	3	5	4
5	8	4	2	3	9	7	6	1
3	7	1	6	5	4	2	8	9
6	5	7	8	4	1	9	2	3
1	4	9	5	2	3	6	7	8
8	2	3	9	7	6	1	4	5
7	9	8	3	6	5	4	1	2
2	1	6	4	9	8	5	3	7
4	3	5	7	1	2	8	9	6

SOLUTION 224

1	3	2	7	9	4	5	8	6
5	6	8	3	1	2	9	4	7
7	4	9	8	5	6	2	3	1
2	1	4	6	3	8	7	5	9
6	7	3	5	2	9	4	1	8
8	9	5	1	4	7	6	2	3
9	8	1	4	7	5	3	6	2
3	5	7	2	6	1	8	9	4
4	2	6	9	8	3	1	7	5

SOLUTION 225

5	2	6	3	8	7	4	9	1
1	4	8	5	2	9	6	3	7
7	9	3	6	1	4	5	8	2
6	5	2	4	9	3	7	1	8
9	1	4	8	7	6	3	2	5
8	3	7	2	5	1	9	4	6
3	8	5	7	4	2	1	6	9
2	6	1	9	3	5	8	7	4
4	7	9	1	6	8	2	5	3

SOLUTION 226

7	5	6	2	4	1	9	8	3
2	9	1	3	6	8	7	4	5
3	4	8	5	9	7	6	1	2
5	3	7	6	8	4	2	9	1
8	2	9	7	1	5	4	3	6
1	6	4	9	2	3	5	7	8
6	1	5	4	3	9	8	2	7
9	7	3	8	5	2	1	6	4
4	8	2	1	7	6	3	5	9

SOLUTION 227

9	1	3	7	5	6	8	2	4
6	8	2	3	1	4	5	9	7
5	4	7	8	9	2	6	1	3
8	6	1	5	2	3	4	7	9
3	5	9	1	4	7	2	6	8
2	7	4	9	6	8	1	3	5
4	3	6	2	7	5	9	8	1
7	9	5	6	8	1	3	4	2
1	2	8	4	3	9	7	5	6

SOLUTION 228

9	5	3	1	6	8	4	7	2
7	4	1	2	5	3	6	9	8
2	6	8	4	9	7	3	5	1
4	1	2	7	8	6	9	3	5
3	8	9	5	4	1	7	2	6
5	7	6	3	2	9	1	8	4
1	3	5	6	7	2	8	4	9
8	2	7	9	1	4	5	6	3
6	9	4	8	3	5	2	1	7

SOLUTION 229

7	4	8	1	5	3	9	6	2
3	5	6	7	9	2	8	1	4
2	9	1	4	6	8	3	7	5
4	7	9	5	8	6	1	2	3
6	3	2	9	1	4	5	8	7
1	8	5	2	3	7	6	4	9
5	1	4	6	7	9	2	3	8
8	6	7	3	2	5	4	9	1
9	2	3	8	4	1	7	5	6

SOLUTION 230

4	2	3	8	1	6	5	9	7
1	5	7	3	9	4	6	8	2
6	9	8	2	5	7	1	4	3
2	8	5	4	3	9	7	6	1
7	6	1	5	2	8	9	3	4
3	4	9	6	7	1	8	2	5
5	7	6	9	4	3	2	1	8
8	1	4	7	6	2	3	5	9
9	3	2	1	8	5	4	7	6

SOLUTION 231

5	2	1	4	8	6	7	3	9
9	3	6	5	7	1	4	8	2
8	4	7	2	3	9	5	6	1
6	5	4	1	9	2	8	7	3
2	1	8	3	6	7	9	5	4
3	7	9	8	4	5	1	2	6
1	6	3	9	5	8	2	4	7
4	8	2	7	1	3	6	9	5
7	9	5	6	2	4	3	1	8

SOLUTION 232

2	6	7	4	8	1	9	3	5
4	1	3	5	6	9	7	8	2
5	9	8	7	3	2	1	6	4
3	5	2	9	7	4	8	1	6
8	4	1	3	5	6	2	9	7
9	7	6	2	1	8	4	5	3
6	2	9	8	4	5	3	7	1
7	8	5	1	2	3	6	4	9
1	3	4	6	9	7	5	2	8

SOLUTION 233

7	1	8	3	4	9	6	5	2
3	6	5	1	2	8	4	9	7
4	2	9	5	6	7	8	3	1
1	9	2	4	8	3	7	6	5
6	3	7	2	1	5	9	4	8
5	8	4	7	9	6	1	2	3
8	5	6	9	7	2	3	1	4
2	7	1	6	3	4	5	8	9
9	4	3	8	5	1	2	7	6

SOLUTION 234

8	3	1	7	4	2	9	5	6
4	2	5	8	9	6	7	1	3
7	6	9	5	1	3	8	4	2
9	4	3	2	7	8	1	6	5
6	5	2	9	3	1	4	8	7
1	8	7	4	6	5	2	3	9
3	9	4	6	8	7	5	2	1
2	7	6	1	5	4	3	9	8
5	1	8	3	2	9	6	7	4

SOLUTION 235

6	1	3	7	2	4	8	5	9
8	4	7	5	9	1	3	6	2
9	5	2	8	3	6	1	7	4
4	9	1	2	6	8	5	3	7
7	2	5	1	4	3	6	9	8
3	6	8	9	5	7	4	2	1
1	3	6	4	7	9	2	8	5
2	7	4	6	8	5	9	1	3
5	8	9	3	1	2	7	4	6

SOLUTION 236

4	2	1	6	5	8	7	9	3
7	9	5	3	4	1	2	8	6
6	3	8	9	7	2	1	4	5
5	4	2	7	3	6	9	1	8
9	7	6	1	8	4	5	3	2
8	1	3	2	9	5	4	6	7
2	8	4	5	6	9	3	7	1
3	5	9	8	1	7	6	2	4
1	6	7	4	2	3	8	5	9

SOLUTION 237

8	2	5	3	1	6	7	4	9
7	3	1	4	9	5	6	2	8
4	6	9	7	2	8	3	5	1
2	5	4	6	7	9	8	1	3
9	8	7	2	3	1	5	6	4
3	1	6	8	5	4	2	9	7
5	4	8	1	6	3	9	7	2
6	7	3	9	4	2	1	8	5
1	9	2	5	8	7	4	3	6

SOLUTION 238

3	2	6	9	4	5	8	1	7
4	7	8	2	3	1	9	5	6
1	5	9	8	7	6	2	3	4
7	3	1	5	6	9	4	2	8
8	4	5	7	2	3	6	9	1
6	9	2	4	1	8	5	7	3
5	1	3	6	9	4	7	8	2
2	8	4	1	5	7	3	6	9
9	6	7	3	8	2	1	4	5

SOLUTION 239

7	8	5	6	4	2	9	1	3
1	4	9	5	7	3	2	6	8
2	6	3	1	9	8	4	7	5
9	3	2	8	6	1	5	4	7
5	7	8	9	2	4	1	3	6
4	1	6	7	3	5	8	2	9
8	2	1	3	5	7	6	9	4
6	5	7	4	1	9	3	8	2
3	9	4	2	8	6	7	5	1

SOLUTION 240

9	4	8	3	2	1	7	5	6
1	2	5	7	6	4	3	9	8
3	6	7	5	8	9	2	1	4
5	7	1	6	3	2	4	8	9
4	8	9	1	5	7	6	3	2
6	3	2	4	9	8	1	7	5
8	1	3	2	4	5	9	6	7
7	5	4	9	1	6	8	2	3
2	9	6	8	7	3	5	4	1

SOLUTION 241

8	1	3	9	2	7	4	5	6
7	5	6	1	8	4	3	9	2
9	2	4	3	5	6	7	1	8
3	4	2	7	1	8	5	6	9
5	9	7	6	4	2	1	8	3
1	6	8	5	3	9	2	7	4
4	7	9	2	6	1	8	3	5
2	3	1	8	9	5	6	4	7
6	8	5	4	7	3	9	2	1

SOLUTION 242

3	4	8	1	6	2	7	9	5
5	2	7	4	8	9	3	6	1
9	1	6	7	5	3	2	8	4
4	3	2	6	9	7	1	5	8
8	9	1	3	4	5	6	2	7
7	6	5	2	1	8	4	3	9
2	5	4	8	3	1	9	7	6
6	7	9	5	2	4	8	1	3
1	8	3	9	7	6	5	4	2

SOLUTION 243

7	2	1	9	8	5	4	6	3
9	4	8	1	3	6	2	7	5
5	6	3	7	4	2	1	9	8
8	3	5	2	7	9	6	1	4
2	9	4	6	1	3	5	8	7
1	7	6	8	5	4	3	2	9
3	8	9	4	6	1	7	5	2
4	1	7	5	2	8	9	3	6
6	5	2	3	9	7	8	4	1

SOLUTION 244

8	7	4	1	5	6	3	9	2
2	6	3	7	4	9	1	8	5
5	9	1	3	8	2	7	4	6
9	3	8	4	2	1	5	6	7
6	5	2	8	3	7	9	1	4
4	1	7	9	6	5	2	3	8
3	8	9	2	7	4	6	5	1
1	2	5	6	9	8	4	7	3
7	4	6	5	1	3	8	2	9

SOLUTION 245

5	1	8	7	3	6	2	9	4
7	4	3	5	2	9	6	1	8
2	6	9	1	4	8	3	5	7
4	5	2	8	1	3	7	6	9
8	3	7	6	9	5	4	2	1
1	9	6	4	7	2	5	8	3
6	2	1	3	8	7	9	4	5
9	7	4	2	5	1	8	3	6
3	8	5	9	6	4	1	7	2

SOLUTION 246

9	5	8	4	2	1	3	7	6
4	1	6	9	3	7	8	5	2
2	7	3	6	5	8	4	9	1
3	6	2	8	7	4	9	1	5
1	9	7	5	6	3	2	8	4
5	8	4	1	9	2	7	6	3
6	2	9	3	8	5	1	4	7
8	3	1	7	4	6	5	2	9
7	4	5	2	1	9	6	3	8

SOLUTION 247

8	1	4	7	9	2	6	3	5
6	5	2	1	4	3	9	7	8
3	7	9	8	6	5	4	1	2
4	8	3	2	5	1	7	6	9
9	6	1	4	7	8	2	5	3
7	2	5	6	3	9	1	8	4
1	4	8	5	2	6	3	9	7
2	9	6	3	8	7	5	4	1
5	3	7	9	1	4	8	2	6

SOLUTION 248

4	1	6	5	9	3	7	8	2
3	5	9	7	2	8	6	4	1
2	7	8	1	6	4	5	3	9
1	4	3	6	5	2	9	7	8
7	6	2	4	8	9	1	5	3
8	9	5	3	1	7	4	2	6
9	2	1	8	7	5	3	6	4
5	8	4	9	3	6	2	1	7
6	3	7	2	4	1	8	9	5

SOLUTION 249

3	5	9	7	8	2	4	1	6
4	2	8	1	3	6	5	9	7
7	1	6	9	4	5	8	3	2
1	3	2	4	6	7	9	5	8
6	9	4	3	5	8	2	7	1
5	8	7	2	9	1	6	4	3
2	6	1	5	7	4	3	8	9
8	4	3	6	1	9	7	2	5
9	7	5	8	2	3	1	6	4

SOLUTION 250

4	5	3	2	8	6	9	1	7
7	1	2	4	5	9	8	3	6
9	6	8	1	3	7	5	4	2
6	4	5	3	1	8	2	7	9
2	9	1	5	7	4	6	8	3
3	8	7	9	6	2	4	5	1
5	3	6	8	9	1	7	2	4
8	2	9	7	4	3	1	6	5
1	7	4	6	2	5	3	9	8

SOLUTION 251

1	6	4	2	8	5	3	9	7
7	3	9	6	4	1	5	2	8
8	5	2	9	7	3	1	6	4
3	2	7	5	6	9	4	8	1
9	8	5	3	1	4	2	7	6
6	4	1	8	2	7	9	3	5
2	9	8	1	5	6	7	4	3
5	7	6	4	3	2	8	1	9
4	1	3	7	9	8	6	5	2

SOLUTION 252

2	8	1	5	7	9	4	3	6
6	5	3	1	4	8	9	2	7
4	7	9	6	2	3	5	8	1
5	6	8	3	1	7	2	9	4
7	3	2	8	9	4	6	1	5
1	9	4	2	5	6	3	7	8
9	2	5	7	6	1	8	4	3
8	1	6	4	3	2	7	5	9
3	4	7	9	8	5	1	6	2

SOLUTION 253

9	3	4	1	6	5	7	2	8
6	7	1	9	2	8	3	5	4
5	2	8	3	4	7	9	1	6
8	9	6	7	1	2	5	4	3
3	1	2	6	5	4	8	7	9
4	5	7	8	3	9	1	6	2
2	4	9	5	8	1	6	3	7
1	8	3	4	7	6	2	9	5
7	6	5	2	9	3	4	8	1

SOLUTION 254

4	5	6	1	7	2	9	8	3
7	3	8	4	6	9	5	1	2
9	1	2	5	8	3	7	4	6
2	4	1	9	3	5	8	6	7
8	7	5	6	2	1	4	3	9
3	6	9	7	4	8	1	2	5
1	8	7	2	9	6	3	5	4
5	2	4	3	1	7	6	9	8
6	9	3	8	5	4	2	7	1

SOLUTION 255

2	5	3	4	9	6	7	8	1
8	1	7	5	3	2	6	9	4
9	6	4	7	1	8	3	5	2
4	2	6	8	5	9	1	3	7
1	7	5	3	2	4	9	6	8
3	8	9	6	7	1	4	2	5
7	4	8	9	6	5	2	1	3
5	9	2	1	4	3	8	7	6
6	3	1	2	8	7	5	4	9

SOLUTION 256

4	2	1	6	8	3	9	7	5
3	8	9	1	5	7	4	2	6
5	7	6	4	2	9	8	1	3
7	5	8	3	1	2	6	4	9
9	6	4	8	7	5	1	3	2
1	3	2	9	4	6	7	5	8
8	1	5	2	6	4	3	9	7
6	9	7	5	3	1	2	8	4
2	4	3	7	9	8	5	6	1

SOLUTION 257

6	3	4	9	1	7	5	8	2
2	1	9	3	5	8	4	6	7
7	8	5	4	6	2	3	1	9
1	5	6	7	3	4	9	2	8
4	7	2	1	8	9	6	5	3
3	9	8	6	2	5	1	7	4
9	2	3	5	7	1	8	4	6
5	6	7	8	4	3	2	9	1
8	4	1	2	9	6	7	3	5

SOLUTION 258

1	9	3	8	6	4	7	2	5
7	4	5	1	2	3	8	6	9
6	2	8	5	7	9	3	4	1
5	7	9	2	4	8	6	1	3
8	1	4	7	3	6	5	9	2
3	6	2	9	5	1	4	8	7
2	5	6	4	1	7	9	3	8
4	8	7	3	9	2	1	5	6
9	3	1	6	8	5	2	7	4

SOLUTION 259

3	5	8	6	2	4	7	1	9
2	4	9	3	7	1	8	5	6
1	6	7	5	8	9	3	2	4
7	3	2	4	9	8	1	6	5
4	9	5	1	3	6	2	8	7
8	1	6	7	5	2	9	4	3
9	7	1	2	6	5	4	3	8
5	8	4	9	1	3	6	7	2
6	2	3	8	4	7	5	9	1

SOLUTION 260

5	2	8	6	3	1	7	9	4
6	7	3	2	4	9	5	8	1
9	1	4	8	7	5	3	2	6
1	4	2	9	8	3	6	5	7
3	6	7	4	5	2	8	1	9
8	5	9	1	6	7	2	4	3
7	8	6	5	1	4	9	3	2
2	3	1	7	9	8	4	6	5
4	9	5	3	2	6	1	7	8

SOLUTION 261

9	6	4	3	2	5	8	1	7
1	2	7	8	6	4	5	3	9
5	8	3	1	9	7	4	6	2
8	7	6	2	3	9	1	4	5
4	3	1	7	5	8	9	2	6
2	5	9	4	1	6	7	8	3
7	1	5	6	4	2	3	9	8
3	9	2	5	8	1	6	7	4
6	4	8	9	7	3	2	5	1

SOLUTION 262

5	3	4	8	9	6	7	1	2
7	9	6	3	1	2	5	8	4
8	2	1	7	5	4	6	3	9
2	1	5	9	3	8	4	7	6
6	7	9	5	4	1	8	2	3
3	4	8	2	6	7	9	5	1
4	8	7	6	2	3	1	9	5
9	6	2	1	7	5	3	4	8
1	5	3	4	8	9	2	6	7

SOLUTION 263

1	6	3	9	4	8	5	2	7
7	4	5	6	3	2	1	9	8
9	2	8	5	7	1	3	4	6
6	3	4	8	9	5	7	1	2
5	9	2	1	6	7	4	8	3
8	7	1	4	2	3	6	5	9
3	8	6	2	1	4	9	7	5
4	5	9	7	8	6	2	3	1
2	1	7	3	5	9	8	6	4

SOLUTION 264

1	8	5	3	4	7	6	2	9
7	4	3	2	9	6	5	1	8
6	2	9	8	5	1	3	7	4
3	1	7	9	8	2	4	6	5
5	6	2	7	3	4	9	8	1
4	9	8	1	6	5	2	3	7
8	7	6	5	2	9	1	4	3
2	5	1	4	7	3	8	9	6
9	3	4	6	1	8	7	5	2

SOLUTION 265

1	4	2	8	5	7	3	9	6
7	8	5	3	9	6	1	4	2
6	3	9	4	2	1	7	8	5
3	2	4	9	6	5	8	7	1
5	6	7	1	4	8	9	2	3
9	1	8	2	7	3	6	5	4
4	7	1	6	8	2	5	3	9
2	5	3	7	1	9	4	6	8
8	9	6	5	3	4	2	1	7

SOLUTION 266

1	7	2	4	5	9	8	6	3
8	5	3	2	7	6	1	9	4
4	6	9	1	8	3	5	7	2
3	8	7	6	2	5	9	4	1
6	4	1	3	9	7	2	8	5
9	2	5	8	1	4	6	3	7
5	3	4	9	6	2	7	1	8
7	9	8	5	4	1	3	2	6
2	1	6	7	3	8	4	5	9

SOLUTION 267

7	8	6	3	4	1	5	2	9
4	9	1	6	2	5	7	3	8
3	2	5	9	8	7	1	6	4
5	6	4	7	1	9	2	8	3
8	3	9	2	5	6	4	7	1
1	7	2	8	3	4	6	9	5
6	4	3	1	9	2	8	5	7
9	5	7	4	6	8	3	1	2
2	1	8	5	7	3	9	4	6

SOLUTION 268

6	1	2	3	7	4	9	8	5
4	7	9	5	6	8	3	1	2
3	8	5	1	9	2	6	7	4
9	6	4	8	5	7	2	3	1
2	5	8	4	3	1	7	6	9
7	3	1	9	2	6	4	5	8
8	2	3	6	4	5	1	9	7
5	9	7	2	1	3	8	4	6
1	4	6	7	8	9	5	2	3

SOLUTION 269

2	3	1	4	9	8	6	7	5
4	5	8	2	7	6	9	3	1
7	6	9	1	3	5	2	4	8
9	2	6	5	8	7	4	1	3
1	4	7	9	2	3	5	8	6
5	8	3	6	1	4	7	9	2
8	7	4	3	6	2	1	5	9
6	1	5	8	4	9	3	2	7
3	9	2	7	5	1	8	6	4

SOLUTION 270

6	8	9	7	3	2	1	4	5
2	7	1	4	6	5	8	9	3
5	3	4	9	1	8	6	2	7
4	5	3	6	9	7	2	1	8
7	6	2	5	8	1	4	3	9
1	9	8	3	2	4	7	5	6
3	4	6	1	7	9	5	8	2
9	2	5	8	4	6	3	7	1
8	1	7	2	5	3	9	6	4

SOLUTION 271

2	1	5	8	6	3	9	7	4
9	7	4	2	1	5	3	6	8
6	8	3	4	7	9	5	1	2
5	6	1	9	2	8	4	3	7
3	2	7	6	5	4	8	9	1
8	4	9	7	3	1	2	5	6
1	5	2	3	4	7	6	8	9
4	3	8	1	9	6	7	2	5
7	9	6	5	8	2	1	4	3

SOLUTION 272

9	4	3	1	2	7	6	8	5
6	2	7	4	8	5	3	9	1
5	8	1	3	6	9	2	7	4
2	3	5	9	4	1	8	6	7
1	6	4	2	7	8	5	3	9
8	7	9	5	3	6	4	1	2
4	5	6	7	9	3	1	2	8
7	1	8	6	5	2	9	4	3
3	9	2	8	1	4	7	5	6

SOLUTION 273

2	1	9	6	8	3	4	5	7
6	3	5	7	9	4	2	1	8
8	7	4	5	2	1	6	9	3
1	4	7	9	3	2	8	6	5
9	6	3	1	5	8	7	2	4
5	8	2	4	7	6	1	3	9
3	2	1	8	4	9	5	7	6
7	9	8	2	6	5	3	4	1
4	5	6	3	1	7	9	8	2

SOLUTION 274

1	3	4	6	5	9	7	2	8
8	7	5	3	2	4	9	1	6
6	2	9	1	8	7	5	3	4
3	5	2	8	9	6	1	4	7
4	8	1	5	7	3	6	9	2
9	6	7	4	1	2	3	8	5
7	1	6	2	3	8	4	5	9
5	9	8	7	4	1	2	6	3
2	4	3	9	6	5	8	7	1

SOLUTION 275

1	7	5	8	4	9	2	3	6
8	9	2	5	3	6	7	4	1
3	4	6	2	1	7	9	8	5
5	3	4	1	2	8	6	7	9
2	1	7	6	9	3	8	5	4
9	6	8	4	7	5	1	2	3
6	8	3	7	5	1	4	9	2
4	5	1	9	8	2	3	6	7
7	2	9	3	6	4	5	1	8

SOLUTION 276

2	5	7	4	8	3	6	9	1
6	9	1	2	7	5	4	3	8
4	8	3	6	9	1	2	5	7
7	2	6	9	5	4	1	8	3
3	1	8	7	6	2	5	4	9
5	4	9	1	3	8	7	6	2
9	6	5	3	1	7	8	2	4
1	3	4	8	2	6	9	7	5
8	7	2	5	4	9	3	1	6

SOLUTION 277

2	1	3	7	4	9	5	8	6
7	8	5	6	3	1	4	2	9
6	4	9	2	8	5	7	3	1
3	2	4	5	6	8	1	9	7
5	6	8	9	1	7	2	4	3
9	7	1	4	2	3	6	5	8
4	3	7	1	9	2	8	6	5
1	9	2	8	5	6	3	7	4
8	5	6	3	7	4	9	1	2

SOLUTION 278

8	5	2	6	3	7	1	4	9
9	3	6	1	4	8	7	5	2
1	4	7	9	5	2	8	6	3
5	8	3	4	9	6	2	1	7
2	1	4	3	7	5	9	8	6
6	7	9	2	8	1	5	3	4
4	2	5	7	1	3	6	9	8
7	9	8	5	6	4	3	2	1
3	6	1	8	2	9	4	7	5

SOLUTION 279

5	8	7	2	4	3	9	6	1
2	6	3	5	9	1	4	7	8
1	4	9	6	7	8	3	2	5
3	7	6	8	5	4	1	9	2
4	1	2	3	6	9	5	8	7
9	5	8	1	2	7	6	4	3
7	3	4	9	1	2	8	5	6
6	9	1	7	8	5	2	3	4
8	2	5	4	3	6	7	1	9

SOLUTION 280

1	5	4	7	8	3	2	9	6
3	2	9	5	6	4	8	7	1
8	7	6	9	2	1	5	4	3
7	4	2	6	9	8	1	3	5
5	8	1	3	4	7	9	6	2
6	9	3	1	5	2	4	8	7
2	3	8	4	1	6	7	5	9
4	6	5	2	7	9	3	1	8
9	1	7	8	3	5	6	2	4

SOLUTION 281

1	2	7	5	3	6	4	8	9
9	5	8	7	2	4	1	3	6
3	6	4	8	9	1	5	7	2
2	7	9	1	4	8	3	6	5
4	8	1	6	5	3	2	9	7
5	3	6	9	7	2	8	1	4
7	4	3	2	8	9	6	5	1
6	9	2	3	1	5	7	4	8
8	1	5	4	6	7	9	2	3

SOLUTION 282

3	2	4	6	9	7	1	8	5
7	1	8	4	5	3	2	6	9
5	6	9	1	8	2	3	7	4
6	4	7	3	1	8	5	9	2
1	3	2	9	7	5	6	4	8
8	9	5	2	6	4	7	3	1
9	7	3	5	4	1	8	2	6
2	5	6	8	3	9	4	1	7
4	8	1	7	2	6	9	5	3

SOLUTION 283

3	4	2	5	8	6	1	9	7
1	9	7	4	3	2	5	6	8
8	6	5	9	1	7	4	3	2
6	1	4	7	9	5	8	2	3
2	5	3	6	4	8	9	7	1
9	7	8	1	2	3	6	5	4
7	3	1	8	6	9	2	4	5
5	8	9	2	7	4	3	1	6
4	2	6	3	5	1	7	8	9

SOLUTION 284

8	7	6	2	5	3	9	1	4
3	1	9	6	7	4	8	5	2
5	4	2	9	1	8	7	6	3
6	2	4	5	8	7	1	3	9
1	3	8	4	2	9	5	7	6
7	9	5	3	6	1	4	2	8
9	8	1	7	3	6	2	4	5
2	6	7	8	4	5	3	9	1
4	5	3	1	9	2	6	8	7

SOLUTION 285

2	4	1	8	5	6	3	9	7
9	8	7	2	3	4	5	1	6
5	6	3	1	7	9	4	8	2
7	9	5	4	1	8	2	6	3
1	3	4	5	6	2	9	7	8
6	2	8	7	9	3	1	4	5
8	7	2	9	4	5	6	3	1
4	5	6	3	8	1	7	2	9
3	1	9	6	2	7	8	5	4

SOLUTION 286

6	3	1	9	7	2	4	8	5
4	8	5	3	1	6	9	2	7
9	7	2	4	8	5	1	3	6
3	6	9	7	5	8	2	4	1
1	4	7	2	3	9	5	6	8
2	5	8	1	6	4	7	9	3
7	9	6	5	2	3	8	1	4
5	2	3	8	4	1	6	7	9
8	1	4	6	9	7	3	5	2

SOLUTION 287

1	7	4	8	3	6	5	2	9
8	5	3	9	1	2	4	7	6
9	6	2	4	7	5	8	3	1
6	1	8	3	9	7	2	4	5
3	9	7	5	2	4	6	1	8
4	2	5	1	6	8	7	9	3
7	3	6	2	5	9	1	8	4
2	8	9	6	4	1	3	5	7
5	4	1	7	8	3	9	6	2

SOLUTION 288

9	3	5	4	2	7	8	6	1
4	6	1	3	9	8	7	2	5
8	7	2	1	5	6	4	3	9
2	8	9	7	4	5	6	1	3
3	1	4	6	8	9	5	7	2
7	5	6	2	3	1	9	4	8
5	4	3	9	6	2	1	8	7
6	9	7	8	1	3	2	5	4
1	2	8	5	7	4	3	9	6

SOLUTION 289

3	7	2	5	6	8	1	9	4
1	4	6	9	2	3	5	8	7
8	9	5	4	7	1	3	6	2
7	1	4	2	5	6	8	3	9
9	2	8	1	3	7	4	5	6
6	5	3	8	4	9	2	7	1
2	3	1	7	9	5	6	4	8
5	8	7	6	1	4	9	2	3
4	6	9	3	8	2	7	1	5

SOLUTION 290

5	8	2	4	3	7	1	9	6
6	3	1	9	2	8	4	5	7
4	9	7	1	6	5	3	2	8
2	1	8	3	7	9	6	4	5
3	7	4	6	5	2	9	8	1
9	5	6	8	4	1	7	3	2
8	2	3	7	1	4	5	6	9
1	4	9	5	8	6	2	7	3
7	6	5	2	9	3	8	1	4

SOLUTION 291

5	3	6	4	7	8	1	9	2
2	4	7	1	9	3	5	6	8
9	1	8	2	6	5	3	7	4
4	7	3	5	1	2	6	8	9
8	6	9	3	4	7	2	1	5
1	5	2	6	8	9	7	4	3
3	9	4	7	5	6	8	2	1
6	8	5	9	2	1	4	3	7
7	2	1	8	3	4	9	5	6

SOLUTION 292

2	1	7	3	8	5	6	4	9
4	6	8	2	1	9	7	5	3
5	3	9	4	6	7	8	2	1
8	5	2	7	4	1	9	3	6
9	7	3	5	2	6	4	1	8
6	4	1	8	9	3	2	7	5
7	9	5	6	3	2	1	8	4
3	8	6	1	7	4	5	9	2
1	2	4	9	5	8	3	6	7

SOLUTION 293

9	4	8	3	7	6	2	5	1
7	1	5	4	2	9	6	8	3
2	3	6	8	5	1	7	4	9
5	2	9	7	3	8	1	6	4
1	6	4	2	9	5	3	7	8
8	7	3	1	6	4	5	9	2
3	8	7	6	4	2	9	1	5
4	9	2	5	1	7	8	3	6
6	5	1	9	8	3	4	2	7

SOLUTION 294

8	6	9	3	2	5	4	7	1
3	2	5	7	4	1	8	6	9
4	1	7	9	6	8	3	2	5
9	8	3	2	7	6	5	1	4
6	7	1	5	3	4	9	8	2
2	5	4	1	8	9	7	3	6
5	9	8	6	1	7	2	4	3
1	4	2	8	5	3	6	9	7
7	3	6	4	9	2	1	5	8

SOLUTION 295

5	8	7	2	9	4	3	6	1
3	2	6	5	8	1	7	4	9
1	9	4	3	6	7	2	5	8
9	4	8	7	3	5	1	2	6
2	3	1	9	4	6	8	7	5
6	7	5	1	2	8	9	3	4
8	6	3	4	7	9	5	1	2
7	1	9	6	5	2	4	8	3
4	5	2	8	1	3	6	9	7

SOLUTION 296

4	1	3	5	2	9	8	6	7
5	8	9	7	3	6	2	1	4
7	6	2	8	1	4	5	3	9
3	4	8	6	5	1	7	9	2
1	5	6	2	9	7	3	4	8
2	9	7	3	4	8	1	5	6
9	2	4	1	7	5	6	8	3
8	3	5	9	6	2	4	7	1
6	7	1	4	8	3	9	2	5

SOLUTION 297

2	8	3	5	6	7	4	1	9
4	5	7	8	9	1	3	2	6
6	1	9	4	3	2	8	5	7
3	4	5	2	8	6	9	7	1
9	2	8	1	7	3	6	4	5
1	7	6	9	5	4	2	3	8
7	6	4	3	1	8	5	9	2
5	3	1	6	2	9	7	8	4
8	9	2	7	4	5	1	6	3

SOLUTION 298

3	7	8	2	6	1	5	4	9
4	6	9	7	3	5	8	2	1
2	1	5	9	4	8	3	6	7
5	2	1	3	9	4	6	7	8
6	3	7	8	5	2	1	9	4
9	8	4	6	1	7	2	3	5
8	9	3	5	7	6	4	1	2
1	5	6	4	2	9	7	8	3
7	4	2	1	8	3	9	5	6

SOLUTION 299

7	1	8	2	5	6	4	3	9
2	3	9	8	4	1	5	7	6
4	5	6	3	9	7	2	8	1
9	8	1	7	3	2	6	5	4
3	7	5	9	6	4	8	1	2
6	2	4	5	1	8	3	9	7
5	6	2	1	8	9	7	4	3
1	4	3	6	7	5	9	2	8
8	9	7	4	2	3	1	6	5

SOLUTION 300

8	5	2	7	4	3	6	1	9
1	3	9	8	6	5	2	7	4
7	6	4	2	9	1	5	8	3
9	8	6	3	5	7	1	4	2
4	1	3	6	2	9	8	5	7
2	7	5	1	8	4	3	9	6
3	4	8	9	1	6	7	2	5
5	2	7	4	3	8	9	6	1
6	9	1	5	7	2	4	3	8